Bayesian Analysis of Capture-Recapture Data with Hidden Markov Models

Bayesian Analysis of Capture-Recapture Data with Hidden Markov Models: Theory and Case Studies in R and NIMBLE introduces ecologists and statisticians to a powerful and unifying framework for analyzing capture-recapture data. Hidden Markov models (HMMs) have become a cornerstone in modern population ecology, offering a flexible way to decompose complex processes such as survival, recruitment, and dispersal into simpler building blocks, while explicitly accounting for the fact that we only observe imperfect data rather than the true underlying states. Combined with Bayesian inference, HMMs provide a natural and transparent approach to handle uncertainty, explore model structures, and draw robust conclusions. This book illustrates how to bring these ideas to life using the R package NIMBLE, a fast-developing environment for building and fitting hierarchical models.

Key Features:

- A clear introduction to the principles of Bayesian statistics, HMMs, and the NIMBLE package
- Step-by-step tutorials showing how to implement a wide range of capture-recapture models for open populations
- Fully reproducible examples with data and R code, following a "learning by doing" philosophy
- Case studies drawn from the ecological literature, illustrating how to apply methods to real-world conservation questions
- Practical guidance on model specification, coding strategies, and interpretation of results

Written in an accessible style, this book is designed for ecologists, wildlife biologists, and conservation scientists who already use R and wish to deepen their modelling toolkit, as well as statisticians interested in ecological applications. Beginners will find a self-contained path into Bayesian capture-recapture modelling, while experienced researchers will discover a flexible framework to extend and adapt to their own data and questions.

Olivier Gimenez is a Research Director at the French National Centre for Scientific Research (CNRS), based at the Centre for Functional and Evolutionary Ecology (CEFE) in Montpellier. Trained as a statistician, he works at the interface of ecology, statistical modelling, and the social sciences, with a particular interest in human-wildlife interactions and population ecology. He coordinates several interdisciplinary projects focusing on mammals and their interactions with human activities. He is the founder of the Statistical Ecology Research Network (GDR Ecologie Statistique), a national network dedicated to statistical ecology. For more than 15 years, he has been teaching statistics to ecologists – especially Bayesian statistics over the past decade – to master's and PhD students.

CHAPMAN & HALL/CRC Interdisciplinary Statistics Series

Series editors: B.J.T. Morgan, C.K. Wikle, P.G.M. van der Heijden

Recently Published Titles

Statistical Methods in Psychiatry and Related Field: Longitudinal, Clustered and Other Repeat Measures Data
R. Gueorguieva

Bayesian Disease Mapping: Hierarchical Modeling in Spatial Epidemiology, Third Edition
A.B. Lawson

Flexbile Imputation of Missing Data, Second Edition
S. van Buuren

Compositional Data Analysis in Practice
M. Greenacre

Applied Directional Statistics: Modern Methods and Case Studies
C. Ley and T. Verdebout

Design of Experiments for Generalized Linear Models
K.G. Russell

Model-Based Geostatistics for Global Public Health: Methods and Applications
P.J. Diggle and E. Giorgi

Statistical and Econometric Methods for Transportation Data Analysis, Third Edition
S. Washington, M.G. Karlaftis, F. Mannering and P. Anastasopoulos

Parameter Redundancy and Identifiability
D. Cole

Mendelian Randomization: Methods for Causal Inference Using Genetic Variants, Second Edition
S. Burgess, S. G. Thompson

Bayesian Modelling of Spatio-Temporal Data with R
Sujit Kumar Sahu

Statistical Methods in Epilepsy
Sharon Chiang, Vikram R. Rao, and Marina Vannucci

Model-Based Monitoring and Statistical Control
Kohei Ohtsu and Genshiro Kitagawa. Translated by the O2 Memorial Project.

Bayesian Analysis of Capture-Recapture Data with Hidden Markov Models: Theory and Case Studies in R and NIMBLE
Olivier Gimenez

For more information about this series, please visit: https://www.routledge.com/Chapman-HallCRC-Interdisciplinary-Statistics/book-series/CHINTSTASER

Bayesian Analysis of Capture-Recapture Data with Hidden Markov Models

Theory and Case Studies in R and NIMBLE

Olivier Gimenez

CRC Press
Taylor & Francis Group
Boca Raton London New York

CRC Press is an imprint of the
Taylor & Francis Group, an **informa** business

A CHAPMAN & HALL BOOK

First edition published 2026
by CRC Press
2385 NW Executive Center Drive, Suite 320, Boca Raton, FL 33431

and by CRC Press
4 Park Square, Milton Park, Abingdon, Oxon, OX14 4RN

CRC Press is an imprint of Taylor & Francis Group, LLC

ISBN: 978-1-032-15423-7 (hbk)
ISBN: 978-1-032-15424-4 (pbk)
ISBN: 978-1-003-24409-7 (ebk)

DOI: 10.1201/9781003244097

Typeset in Alegreya-Regular font
by KnowledgeWorks Global Ltd.

Publisher's note: This book has been prepared from camera-ready copy provided by the author.

Contents

List of Figures   xi

List of Tables   xv

Preface   xvii

I   Foundations   **1**

Introduction   2

1   Bayesian statistics & MCMC   **3**
   1.1   Introduction   3
   1.2   Bayes' theorem   3
   1.3   What is the Bayesian approach?   5
   1.4   Approximating posteriors via numerical integration   7
   1.5   Markov chain Monte Carlo (MCMC)   12
      1.5.1   Monte Carlo integration   13
      1.5.2   Markov chains   14
      1.5.3   Metropolis algorithm   15
   1.6   Assessing convergence   23
      1.6.1   Burn-in   23
      1.6.2   Chain length   26
      1.6.3   What if you have issues of convergence? . . .   29
   1.7   Summary   30
   1.8   Suggested reading   31

2   NIMBLE tutorial   **32**
   2.1   Introduction   32
   2.2   What is NIMBLE?   32
   2.3   Getting started   34

2.4   Programming   47
    2.4.1   NIMBLE functions   47
    2.4.2   Calling R/C++ functions   49
    2.4.3   User-defined distributions   51
2.5   Under the hood   54
2.6   MCMC samplers   61
    2.6.1   Default samplers   61
    2.6.2   User-defined samplers   62
2.7   Tips and tricks   66
    2.7.1   Precision vs standard deviation   67
    2.7.2   Indexing   67
    2.7.3   Faster compilation   67
    2.7.4   Updating MCMC chains   68
    2.7.5   Reproducibility   69
    2.7.6   Parallelization   71
    2.7.7   Incomplete and incorrect initialization . . .   73
    2.7.8   Vectorization   75
2.8   Summary   75
2.9   Suggested reading   76

3   Hidden Markov models   **78**
3.1   Introduction   78
3.2   Longitudinal data   78
3.3   A Markov model for longitudinal data   79
    3.3.1   Assumptions   80
    3.3.2   Transition matrix   80
    3.3.3   Initial states   81
    3.3.4   Likelihood   81
    3.3.5   Example   84
3.4   Bayesian formulation   84
3.5   NIMBLE implementation   86
3.6   Hidden Markov models   91
    3.6.1   Capture-recapture data   91
    3.6.2   Observation matrix   93
    3.6.3   Hidden Markov model   94
    3.6.4   Likelihood   95
3.7   Fitting HMM with NIMBLE   96

	3.8	Marginalization	102
	3.8.1	Brute-force approach	102
	3.8.2	Forward algorithm	104
	3.8.3	NIMBLE implementation	107
3.9		Pooled encounter histories	113
3.10		Decoding after marginalization	118
	3.10.1	Theory	118
	3.10.2	Implementation	121
	3.10.3	Compute first, average after	124
	3.10.4	Average first, compute after	126
3.11		Summary	127
3.12		Suggested reading	129

II Transitions — 131

Introduction — 132

4 Alive and dead — 133

4.1	Introduction	133
4.2	The Cormack-Jolly-Seber (CJS) model	133
4.3	Capture-recapture data	134
4.4	Fitting the CJS model to the dipper data with NIMBLE	136
4.5	CJS model derivatives	142
4.6	Model comparison with WAIC	148
4.7	Goodness of fit	150
	4.7.1 Posterior predictive checks	150
	4.7.2 Classical tests	151
	4.7.3 Design considerations	154
4.8	Covariates	154
	4.8.1 Temporal covariates	155
	4.8.2 Individual covariates	168
	4.8.3 Several covariates	173
	4.8.4 Random effects	179
	4.8.5 Individual time-varying covariates	182
4.9	Why Bayes? Incorporate prior information	188
	4.9.1 Prior elicitation	188
	4.9.2 Moment matching	189
4.10	Summary	190
4.11	Suggested reading	191

5   Sites and states   **193**

  5.1   Introduction   193

  5.2   The Arnason-Schwarz (AS) model   193

      5.2.1   Theory   193

  5.3   Fitting the AS model   196

      5.3.1   Geese data   196

      5.3.2   NIMBLE implementation   198

      5.3.3   Goodness of fit   207

  5.4   What if more than two sites?   208

      5.4.1   Dirichlet prior   209

      5.4.2   Multinomial logit   214

  5.5   Sites may be states   217

      5.5.1   Titis data   217

      5.5.2   The AS model for states   220

      5.5.3   NIMBLE implementation   220

  5.6   Issue of local minima   226

  5.7   Uncertainty   230

      5.7.1   Breeding states   230

      5.7.2   Disease states   237

  5.8   Summary   241

  5.9   Suggested reading   241

III   Case studies   **243**

Introduction   **244**

6   Dealing with covariates   **245**

  6.1   Introduction   245

  6.2   Covariate selection with Reversible Jump MCMC . .   245

      6.2.1   Motivation   245

      6.2.2   Model and NIMBLE implementation   247

      6.2.3   Results and interpretation   250

  6.3   Uncertainty in age   254

      6.3.1   Motivation   254

      6.3.2   Model and NIMBLE implementation   255

      6.3.3   Results and interpretation   259

  6.4   Uncertainty in sex   261

      6.4.1   Motivation   261

6.4.2   Model and NIMBLE implementation   261
6.4.3   Results and interpretation   265

7   Addressing model lack of fit   **268**
7.1   Introduction   268
7.2   Accounting for trap-dependence   268
    7.2.1   Motivation   268
    7.2.2   Model and NIMBLE implementation   269
    7.2.3   Results and interpretation   272
7.3   Allowing your Markov models to remember   272
    7.3.1   Motivation   272
    7.3.2   Model and NIMBLE implementation   273
    7.3.3   Results and interpretation   278
7.4   Accommodating individual heterogeneity   280
    7.4.1   Motivation   280
    7.4.2   Model and NIMBLE implementation   280
    7.4.3   Results and interpretation   292

8   Quantifying life history traits   **294**
8.1   Introduction   294
8.2   Assessing age- and cause-specific mortality   294
    8.2.1   Motivation   294
    8.2.2   Model and NIMBLE implementation   295
    8.2.3   Results and interpretation   299
8.3   Quantifying age-specific breeding   299
    8.3.1   Motivation   299
    8.3.2   Model and NIMBLE implementation   300
    8.3.3   Results and interpretation   307
8.4   Estimating stopover duration   309
    8.4.1   Motivation   309
    8.4.2   Model and NIMBLE implementation   309
    8.4.3   Results and interpretation   312

Conclusion   **315**

Bibliography   **317**

Index   **327**

List of Figures

1.1   The distribution beta(a,b) for different values of a and b. Note that for $a = b = 1$, we get the uniform distribution between 0 and 1 in the top left panel. When a and b are equal, the distribution is symmetric, and the bigger a and b, the more peaked the distribution around the mean (the smaller the variance). The expectation (or mean) of a beta(a,b) is $\dfrac{a}{a+b}$.   10

1.2   MCMC article cover. Source: The Journal of Chemical Physics – https://aip.scitation.org/doi/10.1063/1.169911   12

2.1   Logo of the NIMBLE R package designed by Luke Larson.   33

3.1   Graphical representation of the Viterbi algorithm with $\phi = 0.8$ and $p = 0.6$. States are alive $z = 1$ or dead $z = 2$ and observations are non-detected $y = 1$ or detected $y = 2$.   121

3.2   Comparison of the actual sequences of states to the sequences of states decoded with Viterbi and the – average first, compute after – method.   127

3.3   Comparison of the actual sequences of states to the sequences of states decoded with Viterbi and the – compute first, average after method.   128

4.1   Animal individual marking. Top-left: rings (credit: Emannuelle Cam); Top-right: ear-tags (credit: Jessica Cachelou); Bottom left: coat patterns (credit: Fridolin Zimmermann); Bottom right: ADN feces (credit: Alexander Kopatz)   135

4.2   White-throated Dipper (*Cinclus cinclus*). Credit: Gilbert
       Marzolin.   136

5.1   Canada goose *(Branta canadensis)*. Credit: Max Mc-
       Carthy.   196

5.2   The Dirichlet distribution as a prior for $(\psi^{11}, \psi^{12}, \psi^{13})$
       with vector of parameters α. Here all components of
       α are equal which makes the distribution symmetric,
       and its mean is $(1/3, 1/3, 1/3)$. When $\alpha = 1$, the prior
       for the ψ's is uniform (middle panel), unimodal when
       $\alpha = 10$ (right panel) and concentrated in the corners
       (0 and 1) when $\alpha = 0.1$ (left panel).   209

5.3   The distribution gamma(α,θ) for different values of α
       and θ. The shape argument α determines the overall
       shape while the scale parameter θ only affects the scale
       values (compare the values on X- and Y-axes between
       the bottom and upper panels). The exponential and chi-
       square distributions are particular cases of the gamma
       distribution. If the parameter shape is close to zero, the
       gamma is very similar to the exponential (bottom and
       upper left panels). If the parameter shape is large, then
       the gamma is similar to the chi-squared distribution
       (bottom and upper right panels).   213

5.4   Sooty shearwater (*Ardenna grisea*). Credit: John Harri-
       son.   218

5.5   Influence of the choice of initial values on the conver-
       gence to the global minimum of the deviance illustrated
       with simulated data. The black curve is the profile de-
       viance of the probability to move from site 2 to 1. If an
       initial value is picked in the left area, we end up in the
       local minimum while if it is picked in the green area,
       then we get the global minimum which corresponds to
       the maximum likelihood estimate.   228

5.6   A house finch with a heavy infection caused by conjunc-
       tivitis. Credit: Jim Mondok.   238

8.1 Age-specific mortality probabilities for human-related (solid line) and natural (dashed line) causes in female roe deer at Trois-Fontaines, France, from 1985 to 2013. Shaded polygons represent 95% credible intervals. . . 299

8.2 Arrival probability by sex and week. Means and 95% credible intervals (shaded areas) are provided. 313

8.3 Stopover duration by sex and arrival week. Means and 95% credible intervals (shaded areas) are provided. . 313

List of Tables

6.1 Comparison of parameter estimates for the same HMM to account for age uncertainty: our NIMBLE fit vs. Gervasi et al. (2017, Table 3). 260

6.2 Comparison of parameter estimates for the same HMM to account for sex uncertainty: our NIMBLE fit vs. Pradel et al. (2008, Table 5, model B no genetically sexed anchors). 266

7.1 Second-order (memory) estimates of transition probabilities. 279

7.2 Comparison of posterior estimates from NIMBLE with the data-generating values for a finite–mixture HMM. 285

7.3 Comparison of posterior estimates from NIMBLE with the data-generating values for a finite–mixture HMM. 285

7.4 Posterior estimates vs. data-generating values for a finite–mixture HMM, when a marginalized likelihood is used. 292

8.1 Age-specific estimates of breeding probabilities. Comparison of our NIMBLE estimates with Couet et al. (2019). "NR" denotes quantities not reported in the paper. "NB" is for non-breeding and "yoy" for young-of-the-year. Adult female detection on the paper side is a rough time-average read by eye from the authors' Figure 3 (2005–2016). 308

Preface

Why this book?

The hidden Markov modeling (HMM) framework has gained much attention in the ecological literature over the last decade, and has been suggested as a general modeling framework for the demography of plant and animal populations. HMMs are increasingly used to analyze capture-recapture data and estimate key population parameters (e.g., survival, dispersal, or recruitment) with applications in all fields of ecology (check out https://oliviergimenez.github.io/curated-list-of-HMM-CR-papers/). The first objective of this book is to illustrate the flexibility of HMM to decompose complex problems in smaller pieces that are easier to understand, model and analyze.

In parallel, Bayesian statistics is well established and fast expanding in ecology and related disciplines, as it resonates with the scientific reasoning and offers a natural framework for handling uncertainty. The popularity of Bayesian statistics also comes from the availability of free pieces of software (WinBUGS, OpenBUGS, JAGS, Stan) that allow practitioners to code their own analyses. The second objective of this book is to illustrate the use of the R package NIMBLE [de Valpine et al., 2017] to analyze capture-recapture data with HMM in a Bayesian framework. NIMBLE is seen by many as the future of Bayesian statistical ecology to deal with complex models.

An important part of the book consists in case studies from published papers presented in a tutorial style to abide by the "learning by doing" philosophy. The third objective of this book is to provide reproducible analyses with code and data to teach yourself by example.

Who should read this book?

This book is aimed at beginners who're comfortable using R and write basic code, as well as connoisseurs of capture-recapture who'd like to tap into the power of the Bayesian side of statistics. For both audiences, thinking in the HMM framework will help you in confidently building models and make the most of your capture-recapture data.

I won't lie, there is some math in this book. But the equations I use are either simple enough to follow without a strong mathematical background, or they can be skipped without losing the thread. For those with a solid math background, you'll find there isn't much, and most of the derivations (e.g. posterior distributions) are tucked away in the code. I recognize that a certain level of abstraction can be useful, but my target audience is ecologists. I worried that leaning too heavily on formal math would feel like speaking a foreign language, and to be honest, I'm not even sure I could still "speak math" as fluently as I once did.

What will you learn?

The book is divided into three parts. The first part Foundations is aimed at getting you up-to-speed with Bayesian statistics, NIMBLE, and hidden Markov models. The second part Transitions will teach you all about capture-recapture models for open populations, with reproducible R code to ease the learning process. The third part Case studies provides real-world case studies from the scientific literature that you can reproduce using material covered in previous chapters. These problems can either i) be used to cement and deepen your understanding of methods and models, ii) be adapted for your own purpose, or iii) serve as teaching projects. The data and code are available at https: //github.com/oliviergimenez/banana-book/tree/master/appendix. The last chapter closes the book with take-home messages and recommendations.

What won't you learn?

I do not cover Bayesian statistics or even hidden Markov models exhaustively; I provide just what you need to work with capture-recapture data. If you are interested in knowing more about these topics, hopefully the section `Suggested reading` at the end of each chapter will put you in the right direction. There are also a number of important topics specific to capture-recapture that I do not cover, including closed-population capture-recapture models [Williams et al., 2002], spatial capture-recapture models [Royle et al., 2013], integrated population models [Besbeas and Morgan, 2019], semi-Markov models [e.g. Choquet et al., 2011] and continuous-time models [Rushing, 2023]. These models can be treated as HMMs, but for now the usual formulation is just fine. These developments will be the subject of new chapters in a second edition, hopefully.

Prerequisites

This book uses primarily the R package NIMBLE, so you need to install at least R and NIMBLE. A bunch of other R packages are used. You can install them all at once by running:

```r
install.packages(c(
  "bookdown", "coda", "forecast", "ggtern", "gtools",
  "here", "janitor", "magick", "MCMCvis", "nimble",
  "nimbleEcology", "patchwork", "pdftools",
  "RColorBrewer", "sessioninfo", "tidyverse",
  "wesanderson"
))
```

Besides NIMBLE, I also make frequent use of the tidyverse packages in R and RStudio. For piping, I stick to %>% from the magrittr package.

Since R 4.1.0, R has its own native pipe |>. Most of the code here should work with it too, although I have to admit, I haven't really checked.

How this book was written

I wrote this book in RStudio http://www.rstudio.com/ide/ using bookdown http://bookdown.org/. The book website https://oliv iergimenez.github.io/banana-book is hosted with GitHub Pages https://pages.github.com/, and automatically updated after every push by GitHub Actions https://github.com/features/actions. The source is available from GitHub https://oliviergimenez.github.io/banana-book.

The version of the book you're reading was built with R version 4.5.0 (2025-04-11) and the following packages:

package	version	source
bookdown	0.43	CRAN (R 4.5.0)
coda	0.19–4.1	CRAN (R 4.5.0)
forecast	8.24.0	CRAN (R 4.5.0)
ggtern	3.5.0	CRAN (R 4.5.0)
gtools	3.9.5	CRAN (R 4.5.0)
here	1.0.1	CRAN (R 4.5.0)
janitor	2.2.1	CRAN (R 4.5.0)
magick	2.8.7	CRAN (R 4.5.0)
MCMCvis	0.16.3	CRAN (R 4.5.0)
nimble	1.3.0	CRAN (R 4.5.0)
nimbleEcology	0.5.0	CRAN (R 4.5.0)
patchwork	1.3.0	CRAN (R 4.5.0)
pdftools	3.5.0	CRAN (R 4.5.0)
RColorBrewer	1.1–3	CRAN (R 4.5.0)
sessioninfo	1.2.3	CRAN (R 4.5.0)
tidyverse	2.0.0	CRAN (R 4.5.0)
wesanderson	0.3.7	CRAN (R 4.5.0)

About the author

My name is Olivier Gimenez (https://oliviergimenez.github.io/). I am a senior (euphemism for not so young anymore) scientist at the National Centre for Scientific Research (CNRS; https://www.cnrs.fr/en) in the beautiful city of Montpellier, France.

I struggled studying maths, obtained a PhD in applied statistics a long time ago in a galaxy of wine and cheese. I was awarded my habilitation (https://en.wikipedia.org/wiki/Habilitation) in ecology and evolution so that I could stop pretending to understand what my colleagues were talking about. More recently I embarked in sociology studies because hey, why not.

Lost somewhere at the interface of animal ecology, statistical modeling and social sciences, my so-called expertise lies in population dynamics and species distribution modeling to address questions in ecology and conservation biology about the impact of human activities and the wildlife management. I would be nothing without the students and colleagues who are kind enough to bear with me.

You may find me on BlueSky (https://bsky.app/profile/oaggimenez.bsky.social), LinkedIn (https://www.linkedin.com/in/olivier-gimenez-545451115/), GitHub (https://github.com/oliviergimenez), or get in touch by email at olivier.gimenez@cefe.cnrs.fr.

Acknowledgments

Writing a book is quite an adventure, and a lot of people contributed to make this book a reality. I wish to thank:

- Rob Calver, Sherry Thomas, and Vaishali Singh at Chapman and Hall/CRC, Ashraf Reza and Shashi Kumar at KnowledgeWorks Global Ltd.
- Marc Kéry, Rachel McCrea, Byron Morgan, and Etienne Prévost for their positive reviews of the book proposal I sent to Chapman and Hall/CRC, and their constructive comments and suggestions.

- Marc Kéry for his precious pieces of advice on the process of writing.
- Perry de Valpine, Daniel Turek, Chris Paciorek, and Ben Goldstein for the `NIMBLE` and `nimbleEcology` R packages.
- Colleagues who shared their data; See list at https://github.com/oliviergimenez/banana-book#readme
- People who commented, corrected, offered pieces of advice; See list at https://github.com/oliviergimenez/banana-book#readme.
- Yihui Xie for the `bookdown` R package.
- Attendees of the workshops we run in relation to the content of this book (latest edition was in 2023, see https://oliviergimenez.github.io/bayesian-hmm-cr-workshop-valencia/)
- Perry de Valpine, Sarah Cubaynes, Chloé Nater, Maud Quéroué and Daniel Turek for their help with running workshops in relation to the content of this book (2021 edition: https://oliviergimenez.github.io/bayesian-cr-workshop/; 2022 edition: https://oliviergimenez.github.io/hmm-cr-nimble-isec2022-workshop/).
- Ruth King, Steve Brooks, and Byron Morgan for the workshop on Bayesian statistics for ecologists we taught in Cambridge, the book we wrote together [King et al., 2009], and their contribution to statistical ecology.
- Jean-Dominique Lebreton, Roger Pradel, and Rémi Choquet for the workshops on modeling individual histories with state uncertainty we taught over the years, and sharing their science of capture-recapture with me.
- My employer, the Centre National de la Recherche Scientifique (CNRS), and folks at the Centre d'Écologie Fonctionnelle et Évolutive (CEFE): Being a researcher is a wonderful and meaningful profession, even though academia faces growing challenges with more competition, greater precarity, and fewer permanent positions. I am fortunate to work in a supportive environment at CNRS and CEFE, where people cultivate collaboration and a strong sense of community.
- My family, including my mother, my parents-in-law for their kindness and hospitality, my amazing kids and wonderful wife for putting up with me while I was writing this book.

Part I

Foundations

Introduction

This first part Foundations is aimed at getting you up-to-speed with Bayesian statistics, NIMBLE, and hidden Markov models. The code is available at https://github.com/oliviergimenez/banana-book/tree/master/appendix.

1

Bayesian statistics & MCMC

1.1 Introduction

In this first chapter, you will learn what the Bayesian theory is and how you may use it with a simple example. You will also see how to implement simulation algorithms to implement the Bayesian method for more complex analyses. This is not an exhaustive treatment of Bayesian statistics, but you should get what you need to navigate through the rest of the book.

1.2 Bayes' theorem

Let's not wait any longer and jump into it. Bayesian statistics relies on Bayes' theorem (or law, or rule, whatever you prefer), named after Reverend Thomas Bayes. This theorem was published in 1763 two years after Bayes' death thanks to his friend's efforts Richard Price and was independently discovered by Pierre-Simon Laplace [McGrayne, 2011].

As we will see in a minute, Bayes' theorem is all about conditional probabilities, which are somehow tricky to understand. Conditional probability of outcome or event A given event B, which we denote $\Pr(A \mid B)$, is the probability that A occurs, revised by considering the additional information that event B has occurred. For example, a friend of yours rolls a fair dice and asks you the probability that the outcome was a six (event A). Your answer is 1/6 because each side of the dice is equally likely to come up. Now imagine that you're told the number rolled was even (event B) before you answer your friend's question. Because there are only three even numbers, one of which is six, you may revise your answer for the probability that a six was rolled from 1/6 to $\Pr(A \mid B) = 1/3$. The order in which A and B appear is important; make sure you do not confuse $\Pr(A \mid B)$ and $\Pr(B \mid A)$.

Bayes' theorem gives you $\Pr(A \mid B)$ using marginal probabilities $\Pr(A)$ and $\Pr(B)$ and $\Pr(B \mid A)$:

$$\Pr(A \mid B) = \frac{\Pr(B \mid A)\ \Pr(A)}{\Pr(B)}.$$

We talk about a marginal probability when we are interested in the probability of an event "on its own", without any particular condition. For example, $\Pr(A)$ or $\Pr(B)$ are the overall chances of A or B, without taking anything else into account. It is called marginal because if you built a table with all possible combinations (for instance, dice outcomes classified as even/odd and as "six/not six"), then $\Pr(A)$ and $\Pr(B)$ are obtained by adding up the entries of a row or a column, that is, what you read in the margins of the table.

Originally, Bayes' theorem was seen as a way to infer an unknown cause A of a particular effect B, knowing the probability of effect B given cause A. Think, for example, of a situation where a medical diagnosis is needed, with A an unknown disease and B symptoms, the doctor knows Pr(symptoms|disease) and wants to derive Pr(disease|symptoms). This way of reversing $\Pr(B \mid A)$ into $\Pr(A \mid B)$ explains why Bayesian thinking used to be referred to as "inverse probability".

I don't know about you, but I need to think twice for not messing the letters around.

> I find it easier to remember Bayes' theorem written like this:
>
> $$\Pr(\text{hypothesis} \mid \text{data}) = \frac{\Pr(\text{data} \mid \text{hypothesis})\ \Pr(\text{hypothesis})}{\Pr(\text{data})}$$
>
> The *hypothesis* is a working assumption about which you want to learn using *data*. In capture–recapture analyses, the hypothesis might be a parameter like detection probability, or regression parameters in a relationship between survival probability and a covariate (see Chapter 4). Bayes' theorem tells us how to obtain the probability of a hypothesis given the data we have.

This is great because think about it, this is exactly what the scientific method is! We'd like to know how plausible some hypothesis is based on some data we collected, and possibly compare several hypotheses among them. In that respect, the Bayesian reasoning matches the scientific reasoning, which probably explains why the Bayesian framework is so natural for doing and understanding statistics.

You might ask then, why is Bayesian statistics not the default in statistics? Until recently, there were practical problems to implement Bayes' theorem. Recent advances in computational power coupled with the development of new algorithms have led to a great increase in the application of Bayesian methods within the last three decades.

1.3 What is the Bayesian approach?

Typical statistical problems involve estimating a parameter (or several parameters) θ with available data. To do so, you might be more used to the frequentist rather than the Bayesian method. The frequentist approach, and in particular maximum likelihood estimation (MLE), assumes that the parameters are fixed and have unknown values to be estimated. Therefore, classical estimates are generally point estimates of the parameters of interest. In contrast, the Bayesian approach assumes that the parameters are not fixed and have some unknown distribution. A probability distribution is a mathematical expression that gives the probability for a random variable to take particular values. It may be either discrete (e.g., the Bernoulli, Binomial or Poisson distribution) or continuous (e.g., the Gaussian distribution also known as the normal distribution).

The Bayesian approach is based upon the idea that you, as an experimenter, begin with some prior beliefs about the system. Then you collect data and update your prior beliefs on the basis of observations.

These observations might arise from field work, lab work or from the expertise of your esteemed colleagues. This updating process is based upon Bayes' theorem. Loosely, let's say $A = \theta$ and $B = $ data, then Bayes' theorem gives you a way to estimate parameter θ given the data you have:

$$\Pr(\theta \mid \text{data}) = \frac{\Pr(\text{data} \mid \theta) \times \Pr(\theta)}{\Pr(\text{data})}.$$

Let's spend some time going through each quantity in this formula.

On the left-hand side is $\Pr(\theta \mid \text{data})$, the posterior distribution. It represents what you know after having seen the data. This is the basis for inference and clearly what you're after, a distribution, possibly multivariate if you have more than one parameter.

On the right-hand side, there is $\Pr(\text{data} \mid \theta)$, the likelihood. This quantity is the same as in the MLE approach. Yes, the Bayesian and frequentist approaches have the same likelihood at their core, which mostly explains why results often do not differ much. The likelihood captures the information you have in your data, given a model parameterized with θ.

Then we have $\Pr(\theta)$, the prior distribution. This quantity represents what you know before seeing the data. This is the source of much discussion about the Bayesian approach. It may be vague if you don't know anything about θ. Usually, however, you never start from scratch, and you'd like your prior to reflect the information you have (see Section 4.9 for how to accomplish that).

Last, we have $\Pr(\text{data})$, which is sometimes called the average likelihood because it is obtained by integrating the likelihood with respect to the prior $\Pr(\text{data}) = \int \Pr(\text{data} \mid \theta)\,\Pr(\theta)d\theta$ so that the posterior is standardized, that is, it integrates to one for the posterior to be a distribution. The average likelihood is an integral with dimension the number of parameters θ you need to estimate. This quantity is difficult, if not impossible, to calculate in general. This is one of the reasons why the Bayesian method wasn't used until recently, and why we need algorithms to estimate posterior distributions as I illustrate in the next section.

1.4   Approximating posteriors via numerical integration

Let's take an example to illustrate Bayes' theorem. Say we capture, mark and release $n = 57$ animals at the beginning of winter, out of which we recapture $y = 19$ animals alive [we used a similar example in King et al., 2009]. We'd like to estimate winter survival θ. The data are:

```r
y <- 19 # nb of success
n <- 57 # nb of attempts
```

We build our model first. Assuming all animals are independent of each other and have the same survival probability, then y the number of alive animals at the end of the winter is a binomial distribution with n trials and θ the probability of success:

$$y \sim \text{Binomial}(n, \theta) \qquad\qquad [\text{likelihood}]$$

Note that I follow McElreath [2020] and use labels on the right to help remember what each line is about. This likelihood can be visualized in R:

```r
grid <- seq(0, 1, 0.01) # grid of values for survival
likelihood <- dbinom(y, n, grid) # compute binomial likelihood
df <- data.frame(survival = grid, likelihood = likelihood)
df %>%
  ggplot() +
  aes(x = survival, y = likelihood) +
  geom_line(linewidth = 1.5)
```

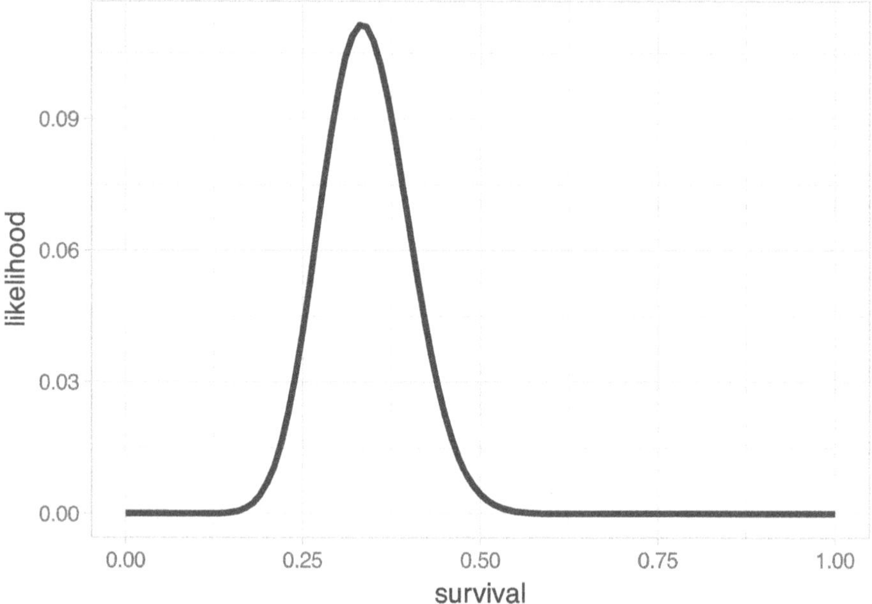

This is the binomial likelihood with $n = 57$ released animals and $y = 19$ survivors after winter. The value of survival (on the x-axis) that corresponds to the maximum of the likelihood function (on the y-axis) is the MLE, or the proportion of success in this example, close to 0.33.

Besides the likelihood, priors are another component of the model in the Bayesian approach. For a parameter that is a probability, the one thing we know is that the prior should be a continuous random variable that lies between 0 and 1. To reflect that, we often go for the uniform distribution $U(0, 1)$ to imply *vague* priors. Here vague means that survival has, before we see the data, the same probability of falling between 0.1 and 0.2 and between 0.8 and 0.9, for example.

$$\theta \sim \text{Uniform}(0, 1) \qquad\qquad [\text{prior for } \theta]$$

Now we apply Bayes' theorem. We write an R function that computes the product of the likelihood times the prior, or the numerator in Bayes' theorem: $\Pr(\text{data} \mid \theta) \times \Pr(\theta)$

```r
numerator <- function(theta) dbinom(y, n, theta) * dunif(theta, 0, 1)
```

We write another function that calculates the denominator, the average likelihood: $\Pr(\text{data}) = \int \Pr(\text{data} \mid \theta)\,\Pr(\theta)d\theta$

```r
denominator <- integrate(numerator,0,1)$value
```

We use the R function `integrate` to calculate the integral in the denominator, which implements quadrature techniques to divide in little squares the area underneath the curve delimited by the function to integrate (here the numerator), and count them.

Then we get a numerical approximation of the posterior of winter survival by applying Bayes' theorem:

```r
grid <- seq(0, 1, 0.01) # grid of values for theta
numerical_posterior <- data.frame(survival = grid,
   posterior = numerator(grid)/denominator) # Bayes' theorem
numerical_posterior %>%
  ggplot() +
  aes(x = survival, y = posterior) +
  geom_line(linewidth = 1.5)
```

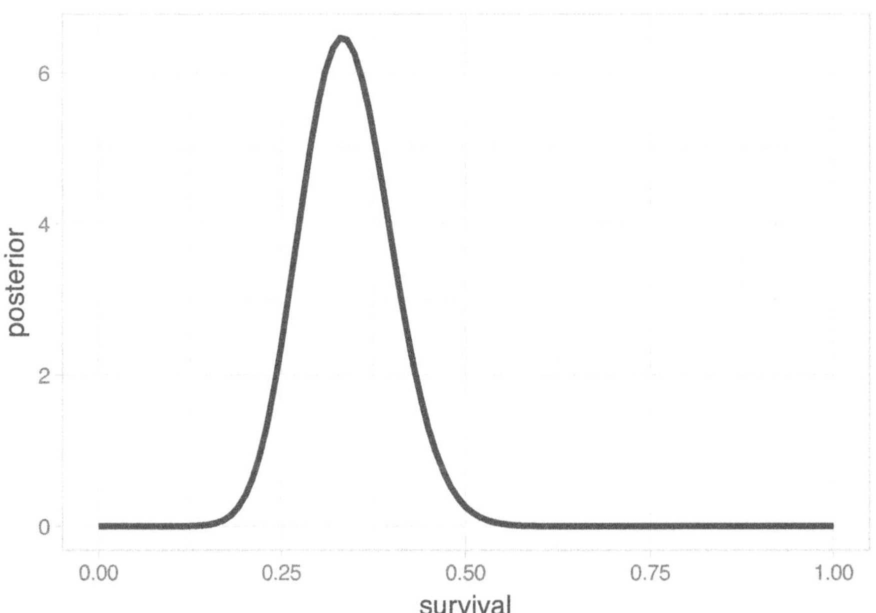

How good is our numerical approximation of the survival posterior distribution? Ideally, we would want to compare the approximation to the true posterior distribution. Although a closed-form expression for the posterior distribution is in general intractable, when you combine a binomial likelihood together with a beta distribution as a prior, then the posterior distribution is also a beta distribution, which makes it amenable to all sorts of exact calculations. We say that the beta distribution is the conjugate prior distribution for the binomial distribution. The beta distribution is continuous between 0 and 1, and extends the uniform distribution to situations where not all outcomes are equally likely. It has two parameters a and b that control its shape (Figure 1.1).

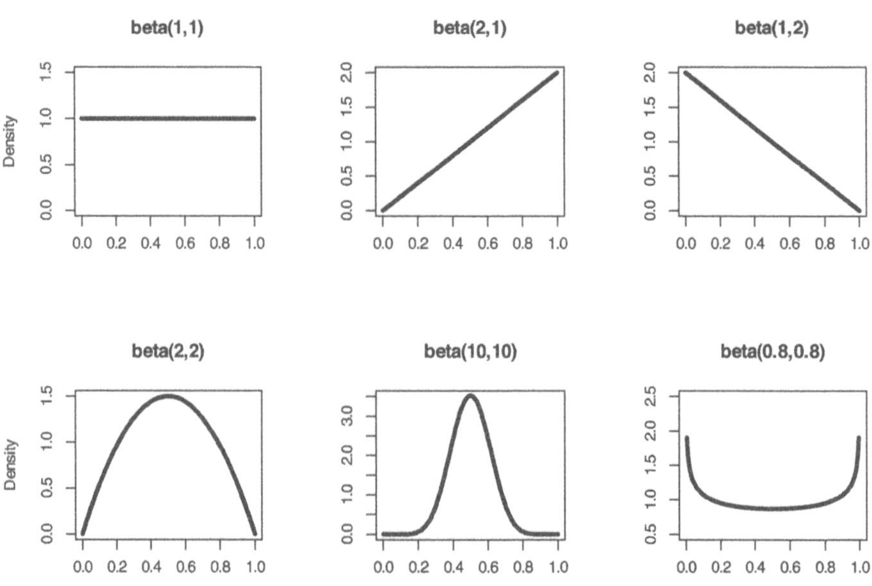

FIGURE 1.1: The distribution beta(a,b) for different values of a and b. Note that for $a = b = 1$, we get the uniform distribution between 0 and 1 in the top left panel. When a and b are equal, the distribution is symmetric, and the bigger a and b, the more peaked the distribution around the mean (the smaller the variance). The expectation (or mean) of a beta(a,b) is $\dfrac{a}{a + b}$.

If the likelihood of the data y is binomial with n trials and probability of success θ, and the prior is a beta distribution with parameters a and b, then the posterior is a beta distribution with parameters

$a + y$ and $b + n - y$. In our example, we have $n = 57$ trials and $y = 19$ animals that survived and a uniform prior between 0 and 1 or a beta distribution with parameters $a = b = 1$, therefore survival has a beta posterior distribution with parameters 20 and 39. Let's superimpose the exact posterior and the numerical approximation:

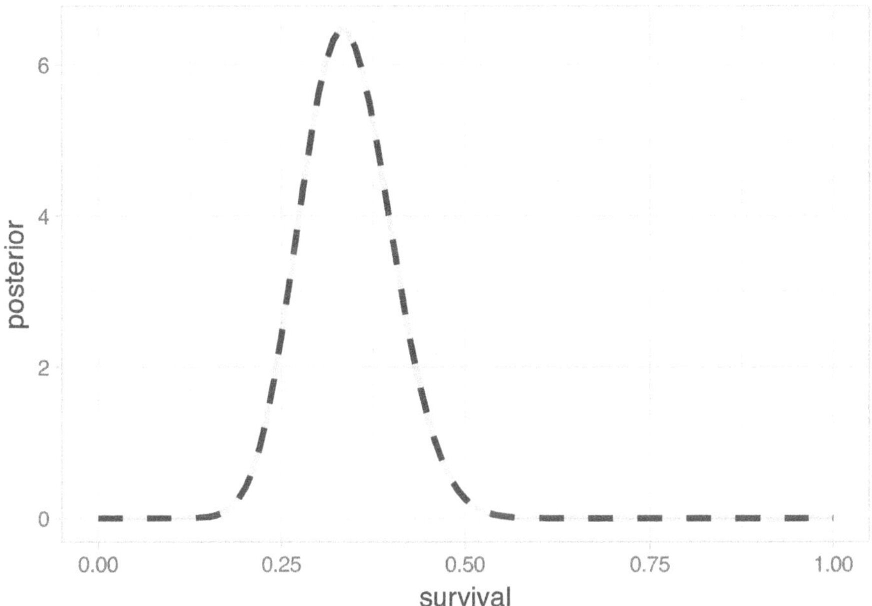

Clearly, the exact (black) vs. numerical approximation (grey) of winter survival posterior distribution is indistinguishable, suggesting that the numerical approximation is more than fine.

In our example, we have a single parameter to estimate, winter survival. This means dealing with a one-dimensional integral in the denominator which is pretty easy with quadrature techniques and the R function `integrate()`. Now what if we had multiple parameters? For example, imagine you'd like to fit a capture-recapture model with detection probability p and regression parameters α and β for the intercept and slope of a relationship between survival probability and a covariate, then Bayes' theorem gives you the posterior distribution of all three parameters together:

$$\Pr(\alpha, \beta, p \mid \text{data}) = \frac{\Pr(\text{data} \mid \alpha, \beta, p) \times \Pr(\alpha, \beta, p)}{\iiint \Pr(\text{data} \mid \alpha, \beta, p)\, \Pr(\alpha, \beta, p)\, d\alpha\, d\beta\, dp}$$

There are two computational challenges with this formula. First, do we really wish to calculate a three-dimensional integral? The answer is no, one-dimensional and two-dimensional integrals are so much further we can go with standard methods. Second, we're more interested in a posterior distribution for each parameter separately than the joint posterior distribution. The so-called marginal distribution of p for example is obtained by integrating over all the other parameters – a two-dimensional integral in this example. Now imagine with tens or hundreds of parameters to estimate, these integrals become highly multi-dimensional and simply intractable. In the next section, I introduce powerful simulation methods to circumvent this issue.

1.5   Markov chain Monte Carlo (MCMC)

In the early 1990s, statisticians rediscovered work from the 1950s in physics. In a famous paper that would lay the foundations of modern Bayesian statistics (Figure 1.2), the authors use simulations to approximate posterior distributions with some precision by drawing large samples. This is a neat trick to avoid explicit calculation of the multi-dimensional integrals we struggle with when using Bayes' theorem.

THE JOURNAL OF CHEMICAL PHYSICS   VOLUME 21, NUMBER 6   JUNE, 1953

Equation of State Calculations by Fast Computing Machines

NICHOLAS METROPOLIS, ARIANNA W. ROSENBLUTH, MARSHALL N. ROSENBLUTH, AND AUGUSTA H. TELLER,
Los Alamos Scientific Laboratory, Los Alamos, New Mexico

AND

EDWARD TELLER,* *Department of Physics, University of Chicago, Chicago, Illinois*
(Received March 6, 1953)

A general method, suitable for fast computing machines, for investigating such properties as equations of state for substances consisting of interacting individual molecules is described. The method consists of a modified Monte Carlo integration over configuration space. Results for the two-dimensional rigid-sphere system have been obtained on the Los Alamos MANIAC and are presented here. These results are compared to the free volume equation of state and to a four-term virial coefficient expansion.

FIGURE 1.2: MCMC article cover. Source: The Journal of Chemical Physics – https://aip.scitation.org/doi/10.1063/1.169911

These simulation algorithms are called Markov chain Monte Carlo (MCMC), and they definitely gave a boost to Bayesian statistics. There are two parts in MCMC, Markov chain and Monte Carlo, let's try and make sense of these terms.

1.5.1 Monte Carlo integration

What does Monte Carlo stand for? Monte Carlo integration is a simulation technique to calculate integrals of any function f of random variable X with distribution $\Pr(X)$ say $\int f(X)\Pr(X)dX$. You draw values $X_1, \ldots, X_k$ from $\Pr(X)$, the distribution of X, apply function f to these values, then calculate the mean of these new values $\dfrac{1}{k}\sum_{i=1}^{k} f(X_i)$ to approximate the integral. How is Monte Carlo integration used in a Bayesian context? The posterior distribution contains all the information we need about the parameter to be estimated. When dealing with many parameters however, you may want to summarize posterior results by calculating numerical summaries. The simplest numerical summary is the mean of the posterior distribution, $E(\theta) = \int \theta \Pr(\theta|\text{data})$, where X is θ now and f is the identity function. Posterior mean can be calculated with Monte Carlo integration:

```r
# draw 1000 values from posterior survival beta(20,39)
sample_from_posterior <- rbeta(1000, 20, 39)
# compute mean with Monte Carlo integration
mean(sample_from_posterior)
## [1] 0.3421
```

You may check that the mean we have just calculated matches closely the expectation of a beta distribution:

```r
20/(20+39) # expectation of beta(20,39)
## [1] 0.339
```

Another useful numerical summary is the credible interval within which our parameter falls with some probability, usually 0.95 hence a 95% credible interval. Finding the bounds of a credible interval requires calculating quantiles, which in turn involves integrals and the use of

Monte Carlo integration. A 95% credible interval for winter survival can be obtained in R with:

```r
quantile(sample_from_posterior, probs = c(2.5/100, 97.5/100))
##   2.5%   97.5%
## 0.2258 0.4671
```

1.5.2 Markov chains

What is a Markov chain? A Markov chain is a random sequence of numbers, in which each number depends only on the previous number. An example is the weather in my home town in Southern France, Montpellier, in which a sunny day is most likely to be followed by another sunny day, say with probability 0.8, and a rainy day is rarely followed by another rainy day, say with probability 0.1. The dynamic of this Markov chain is captured by the transition matrix Γ:

$$\Gamma = \begin{pmatrix} & \text{sunny tomorrow} & \text{rainy tomorrow} \\ 0.8 & 0.2 \\ 0.9 & 0.1 \end{pmatrix} \begin{matrix} \text{sunny today} \\ \text{rainy today} \end{matrix}$$

In rows the weather today, and in columns the weather tomorrow. The cells give the probability of a sunny or rainy day tomorrow, given the day is sunny or rainy today. Under certain conditions, a Markov chain will converge to a unique stationary distribution. In our weather example, let's run the Markov chain for 20 steps:

```r
# transition matrix
weather <- matrix(c(0.8, 0.2, 0.9, 0.1), nrow = 2, byrow = T)
steps <- 20
for (i in 1:steps){
  weather <- weather %*% weather # matrix multiplication
}
round(weather, 2) # matrix product after 20 steps
##        [,1] [,2]
## [1,] 0.82 0.18
## [2,] 0.82 0.18
```

Each row of the transition matrix converges to the same distribution $(0.82, 0.18)$ as the number of steps increases. Convergence happens no matter which state you start in, and you always have probability 0.82 of the day being sunny and 0.18 of the day being rainy.

Back to MCMC, the core idea is that you can build a Markov chain with a given stationary distribution set to be the desired posterior distribution.

> Putting Monte Carlo and Markov chains together, MCMC allows us to generate a sample of values (Markov chain) whose distribution converges to the posterior distribution, and we can use this sample of values to calculate any posterior summaries (Monte Carlo), such as posterior means and credible intervals.

1.5.3 Metropolis algorithm

There are several ways of constructing Markov chains for Bayesian inference. You might have heard about the Metropolis-Hastings or the Gibbs sampler. Have a look to https://chi-feng.github.io/mcmc-demo/ for an interactive gallery of MCMC algorithms. Here I illustrate the Metropolis algorithm and how to implement it in practice.

Let's go back to our example on animal survival estimation. We illustrate sampling from survival posterior distribution. We write functions for likelihood, prior and posterior:

```r
# 19 recaptured alive out of 57 captured, marked and released
survived <- 19
released <- 57


# binomial log-likelihood function
loglikelihood <- function(x, p){
  dbinom(x = x, size = released, prob = p, log = TRUE)
}


# uniform prior density
```

```r
logprior <- function(p){
  dunif(x = p, min = 0, max = 1, log = TRUE)
}


# posterior density function (log scale)
posterior <- function(x, p){
  loglikelihood(x, p) + logprior(p) # - log(Pr(data))
}
```

The Metropolis algorithm works as follows:

1. We pick a value of the parameter to be estimated. This is where we start our Markov chain – this is a *starting* value, or a starting location.

2. To decide where to go next, we propose to move away from the current value of the parameter – this is a *candidate* value. To do so, we add to the current value some random value from e.g. a normal distribution with some variance – this is a *proposal* distribution. The Metropolis algorithm is a particular case of the Metropolis-Hastings algorithm with symmetric proposals.

3. We compute the ratio of the probabilities at the candidate and current locations $R = \dfrac{\Pr(\text{candidate}|\text{data})}{\Pr(\text{current}|\text{data})}$. This is where the magic of MCMC happens, in that $\Pr(\text{data})$, the denominator in the Bayes' theorem, appears in both the numerator and the denominator in R therefore cancels out and does not need to be calculated.

4. If the posterior at the candidate location $\Pr(\text{candidate}|\text{data})$ is higher than at the current location $\Pr(\text{current}|\text{data})$, in other words when the candidate value is more plausible than the current value, we definitely accept the candidate value. If not, then we accept the candidate value with probability R and reject with probability $1 - R$. For example, if the candidate

value is ten times less plausible than the current value, then we accept with probability 0.1 and reject with probability 0.9. How does it work in practice? We use a continuous spinner that lands somewhere between 0 and 1 – call the random spin X. If X is smaller than R, we move to the candidate location, otherwise we remain at the current location. We do not want to accept or reject too often. In practice, the Metropolis algorithm should have an acceptance probability between 0.2 and 0.4, which can be achieved by *tuning* the variance of the normal proposal distribution.

5.   We repeat 2–4 a number of times – or *steps*.

Enough of the theory, let's implement the Metropolis algorithm in R. Let's start by setting the scene:

```r
steps <- 100 # number of steps
theta.post <- rep(NA, steps) # vector to store samples
accept <- rep(NA, steps) # keep track of accept/reject
set.seed(1234) # for reproducibility
```

Now follow the 5 steps we've just described. First, we pick a starting value, and store it (step 1):

```r
inits <- 0.5
theta.post[1] <- inits
accept[1] <- 1
```

Then, we need a function to propose a candidate value:

```r
# by default, standard deviation of the proposal distribution is 1
move <- function(x, away = 1){
  # apply logit transform (-infinity,+infinity)
  logitx <- log(x / (1 - x))
  # add a value taken from N(0,sd=away) to current value
  logit_candidate <- logitx + rnorm(1, 0, away)
```

```r
  # back-transform (0,1)
  candidate <- plogis(logit_candidate)
  return(candidate)
}
```

We add a value taken from a normal distribution with mean zero and standard deviation we call *away*. We work on the logit scale to make sure the candidate value for survival lies between 0 and 1.

Now we're ready for steps 2, 3 and 4. We write a loop to take care of step 5. We start at initial value 0.5 and run the algorithm for 100 steps or iterations:

```r
for (t in 2:steps){ # repeat steps 2-4 (step 5)

  # propose candidate value for survival (step 2)
  theta_star <- move(theta.post[t-1])


  # calculate ratio R (step 3)
  pstar <- posterior(survived, p = theta_star)
  pprev <- posterior(survived, p = theta.post[t-1])
  logR <- pstar - pprev # likelihood and prior are on the log scale
  R <- exp(logR)


  # accept candidate value or keep current value (step 4)
  X <- runif(1, 0, 1) # spin continuous spinner
  if (X < R){
    theta.post[t] <- theta_star # accept candidate value
    accept[t] <- 1 # accept
  }
  else{
    theta.post[t] <- theta.post[t-1] # keep current value
    accept[t] <- 0 # reject
  }
}
```

We get the following values:

```r
head(theta.post) # first values
## [1] 0.5000 0.2302 0.2906 0.2906 0.2980 0.2980
tail(theta.post) # last values
## [1] 0.2622 0.2622 0.2622 0.3727 0.3232 0.3862
```

Visually, you may look at the chain:

```r
df <- data.frame(x = 1:steps, y = theta.post)
df %>%
  ggplot() +
  geom_line(aes(x = x, y = y),
            size = 1.5,
            color = wesanderson::wes_palettes$Zissou1[1]) +
  labs(x = "iterations", y = "samples") +
  ylim(0.1, 0.6)
```

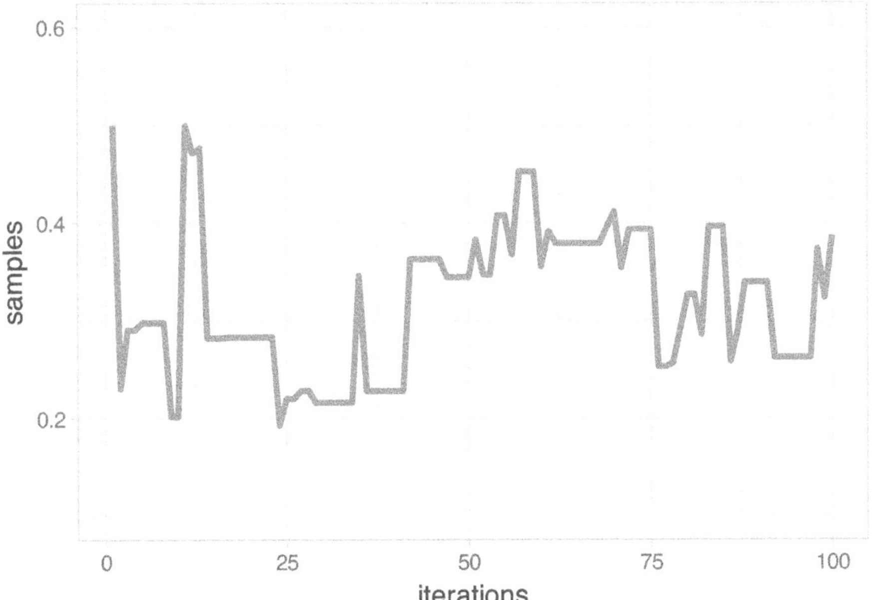

In this visualization, remember that our Markov chain starts at value 0.5. The steps or iterations are on the x-axis, and samples on the y-axis. This graphical representation is called a trace plot.

The acceptance probability is the average number of times we accepted a candidate value, which is 0.44 and almost satisfying.

To make our life easier and avoid repeating the same lines of code again and again, let's make a function out of the code we have written so far:

```r
metropolis <- function(steps = 100, inits = 0.5, away = 1){

  # pre-alloc memory
  theta.post <- rep(NA, steps)


  # start
  theta.post[1] <- inits


  for (t in 2:steps){

    # propose candidate value for prob of success
    theta_star <- move(theta.post[t-1], away = away)


    # calculate ratio R
    pstar <- posterior(survived, p = theta_star)
    pprev <- posterior(survived, p = theta.post[t-1])
    logR <- pstar - pprev
    R <- exp(logR)


    # accept candidate value or keep current value (step 4)
    X <- runif(1, 0, 1) # spin continuous spinner
    if (X < R){
      theta.post[t] <- theta_star
    }
    else{
      theta.post[t] <- theta.post[t-1]
    }
  }
  theta.post
}
```

Can we run another chain and start at initial value 0.2 this time? Yes, just go through the same algorithm again, and visualize the results with trace plot of survival for two chains starting at 0.2 (grey) and 0.5 (black) run for 100 steps:

```r
theta.post2 <- metropolis(steps = 100, inits = 0.2)
df2 <- data.frame(x = 1:steps, y = theta.post2)
ggplot() +
  geom_line(data = df,
            aes(x = x, y = y),
            size = 1.5,
            color = wesanderson::wes_palettes$Zissou1[1]) +
  geom_line(data = df2,
            aes(x = x, y = y),
            size = 1.5,
            color = wesanderson::wes_palettes$Zissou1[3]) +
  labs(x = "iterations", y = "values from posterior distribution") +
  ylim(0.1, 0.6)
```

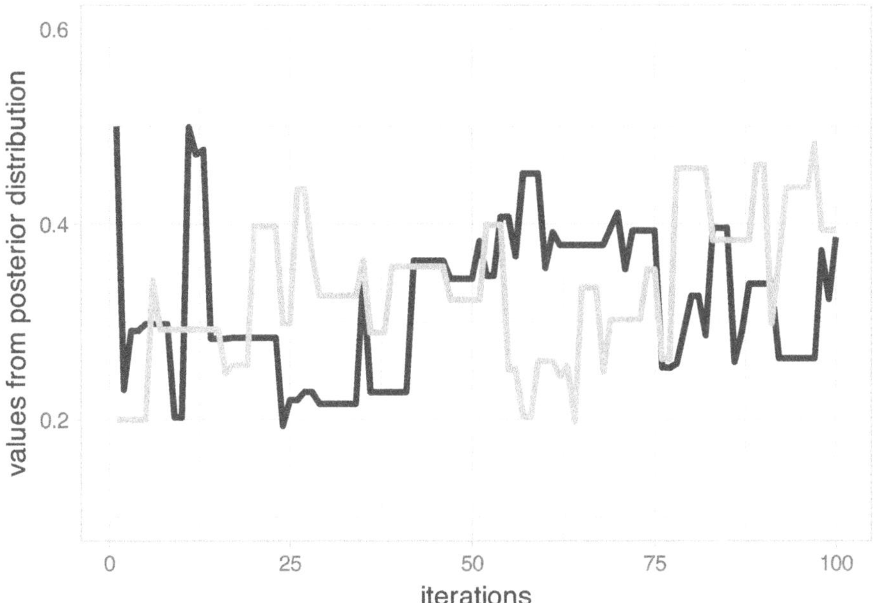

Notice that we do not get the exact same results because the algorithm is stochastic. The question is to know whether we have reached the stationary distribution. Let's increase the number of steps, start at 0.5 and run a chain with 5000 iterations:

```r
steps <- 5000
set.seed(1234)
theta.post <- metropolis(steps = steps, inits = 0.5)
df <- data.frame(x = 1:steps, y = theta.post)
df %>%
  ggplot() +
  geom_line(aes(x = x, y = y),
            size = 1,
            color = wesanderson::wes_palettes$Zissou1[1]) +
  labs(x = "iterations", y = "values from posterior distribution") +
  ylim(0.1, 0.6) +
  geom_hline(aes(yintercept = mean(theta.post),
                 linetype = "posterior mean")) +
  scale_linetype_manual(name = "", values = c(2,2))
```

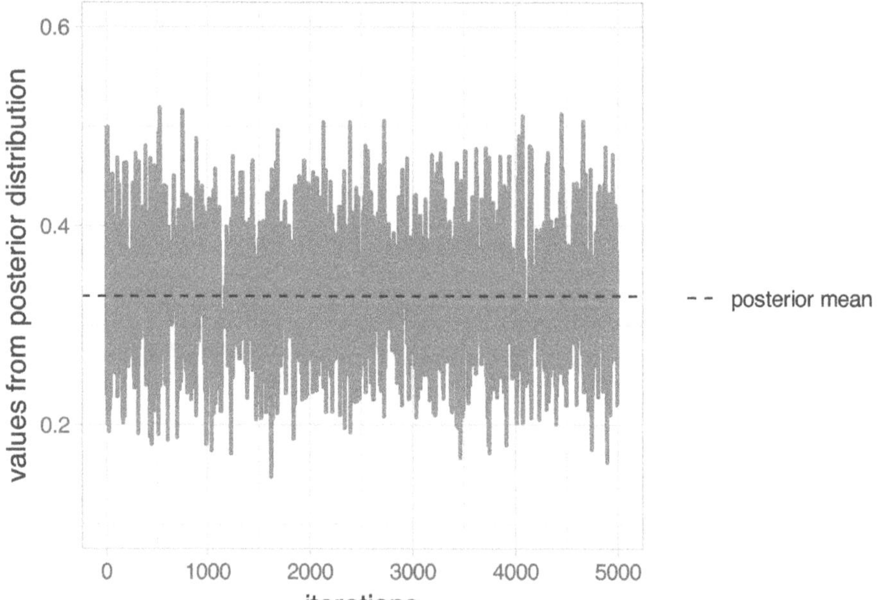

This is what we're after, a trace plot that looks like a beautiful lawn, see Section 1.6.

Once the stationary distribution is reached, you may regard the realizations of the Markov chain as a sample from the posterior distribution, and obtain numerical summaries. In the next section, we consider several important implementation issues.

1.6 Assessing convergence

> When implementing MCMC, we need to determine how long it takes for our Markov chain to converge to the target distribution, and the number of iterations we need after achieving convergence to get reasonable Monte Carlo estimates of numerical summaries (posterior means and credible intervals).

1.6.1 Burn-in

In practice, we discard observations from the start of the Markov chain and just use observations from the chain once it has converged. The initial observations that we discard are usually referred to as the *burn-in*.

The simplest method to determine the length of the burn-in period is to look at trace plots. Going back to our example, let's have a look to a trace plot of a chain that starts at value 0.99.

```r
# set up the scene
steps <- 1000
theta.post <- metropolis(steps = steps, inits = 0.99)
df <- data.frame(x = 1:steps, y = theta.post)
df %>%
  ggplot() +
  geom_line(aes(x = x, y = y),
            size = 1.2,
            color = wesanderson::wes_palettes$Zissou1[1]) +
  labs(x = "iterations", y = "survival") +
```

```r
theme_light(base_size = 14) +
annotate("rect",
         xmin = 0,
         xmax = 100,
         ymin = 0.1,
         ymax = 1,
         alpha = .3) +
scale_y_continuous(expand = c(0,0))
```

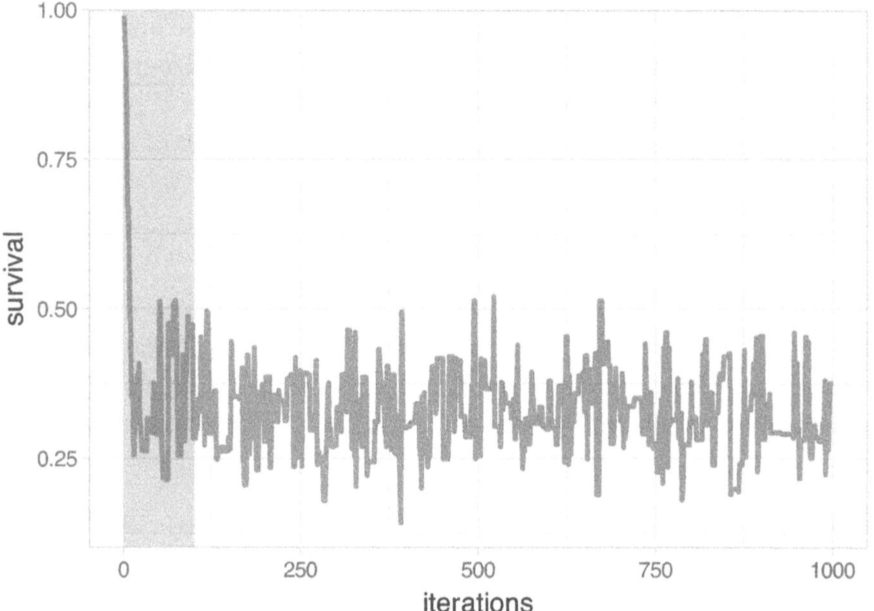

The chain starts at value 0.99 and rapidly stabilizes, with values bouncing back and forth around 0.3 from the 100th iteration onwards. You may choose the shaded area as the burn-in, and discard the first 100th values.

We see from the trace plot below that we need at least 100 iterations to achieve convergence toward an average survival around 0.3. It is always better to be conservative when specifying the length of the burn-in period, and in this example, we would use 250 or even 500 iterations as a burn-in. The length of the burn-in period can be determined by performing preliminary MCMC short runs.

Inspecting the trace plot for a single run of the Markov chain is useful. However, we usually run the Markov chain several times, starting from different over-dispersed points, to check that all runs achieve the same stationary distribution. This approach is formalized by using the Brooks-Gelman-Rubin (BGR) statistic $\hat{R}$ which measures the ratio of the total variability combining multiple chains (between-chain plus within-chain) to the within-chain variability. The BGR statistic asks whether there is a chain effect and is very much alike the F test in an analysis of variance. Values below 1.1 indicate likely convergence.

Back to our example, we run two Markov chains with starting values 0.2 and 0.8 using 100 up to 5000 iterations, and calculate the BGR statistic using half the number of iterations as the length of the burn-in (code not shown):

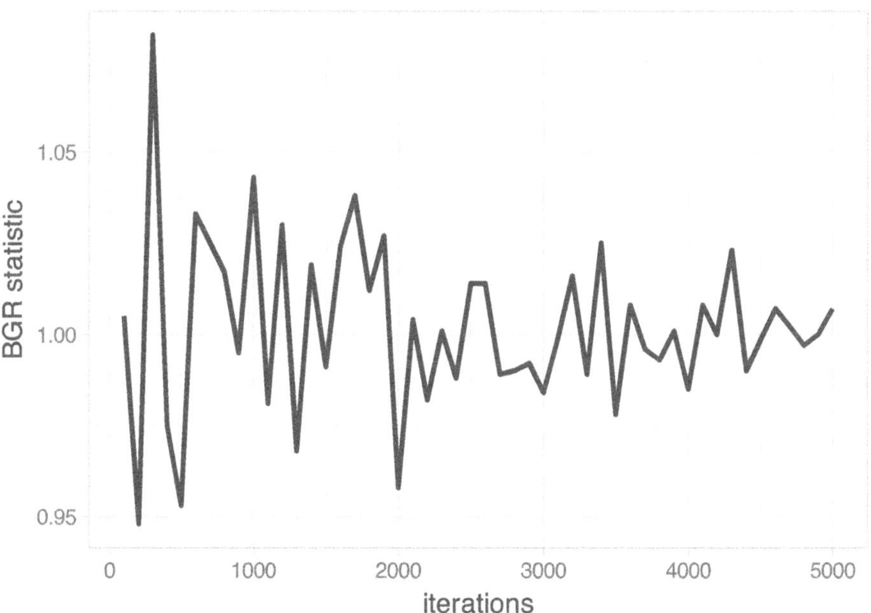

We get a value of the BGR statistic near 1 by up to 2000 iterations, which suggests that with 2000 iterations as a burn-in, there is no evidence of a lack of convergence.

It is important to bear in mind that a value near 1 for the BGR statistic is only a necessary *but not sufficient* condition for convergence. In other

words, this diagnostic cannot tell you for sure that the Markov chain
has achieved convergence, only that it has not.

1.6.2   Chain length

How long of a chain is needed to produce reliable parameter estimates?
To answer this question, you need to keep in mind that successive
steps in a Markov chain are not independent – this is usually referred
to as *autocorrelation*. Ideally, we would like to keep autocorrelation as
low as possible. Here again, trace plots are useful to diagnose issues
with autocorrelation. Let's get back to our survival example. The figure
below shows trace plots for different values of the standard deviation
(parameter *away*) of the normal proposal distribution we use to propose
a candidate value (Section 1.5.3):

```r
# inspired from
# https://bookdown.org/content/3686/markov-chain-monte-carlo.html


n_steps <- 10000


d <-
  tibble(away = c(0.1, 1, 10)) %>%
  mutate(accepted_traj = map(away,
                             metropolis,
                             steps = n_steps,
                             inits = 0.1)) %>%
  unnest(accepted_traj)


d <-
  d %>%
  mutate(proposal_sd = str_c("Proposal SD = ", away),
         iter        = rep(1:n_steps, times = 3))


trace <- d %>%
  ggplot(aes(y = accepted_traj, x = iter)) +
  geom_path(size = 1/4, color = "steelblue") +
  geom_point(size = 1/2, alpha = 1/2, color = "steelblue") +
```

```
scale_y_continuous("survival",
                   breaks = 0:5 * 0.1,
                   limits = c(0.15, 0.5)) +
scale_x_continuous("iterations",
                   breaks = seq(n_steps-n_steps*10/100,
                               n_steps,
                               by = 600),
                   limits = c(n_steps-n_steps*10/100,n_steps)) +
facet_wrap(~proposal_sd, ncol = 3) +
theme_light(base_size = 14)
```

trace

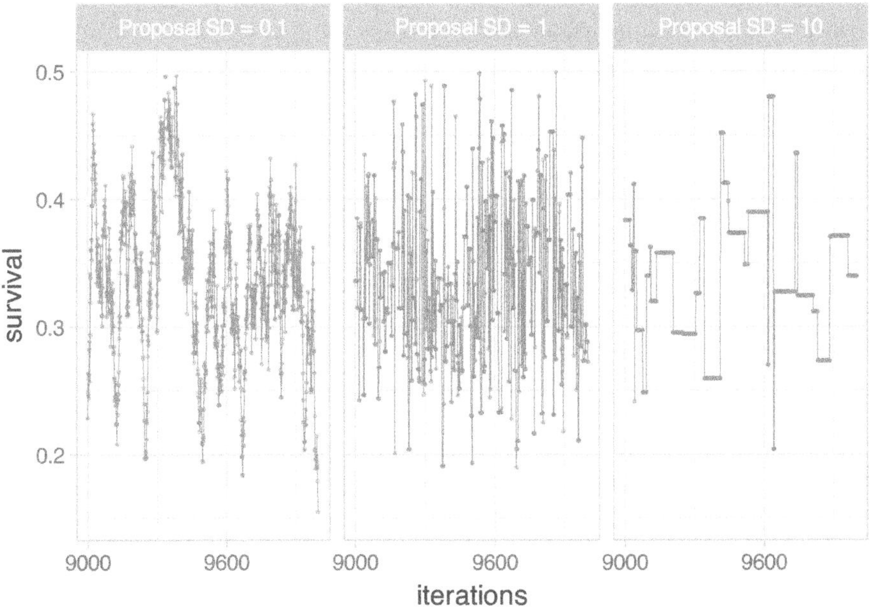

Small and big moves in the left and right panels provide high correlations between successive observations of the Markov chain, whereas a standard deviation of 1 in the center panel allows efficient exploration of the parameter space. The movement around the parameter space is referred to as *mixing*. Mixing is bad when the chain makes small and big moves, and good otherwise.

In addition to trace plots, autocorrelation function (ACF) plots are a convenient way of displaying the strength of autocorrelation in a given sample values. ACF plots provide the autocorrelation between successively sampled values separated by an increasing number of iterations, or *lag*. We obtain the autocorrelation function plots for different values of the standard deviation of the proposal distribution with the R `forecast::ggAcf()` function:

```r
library(forecast)
mask <- d$proposal_sd=="Proposal SD = 0.1"
plot1 <- ggAcf(x = d$accepted_traj[mask]) + ggtitle("SD = 0.1")
mask <- d$proposal_sd=="Proposal SD = 1"
plot2 <- ggAcf(x = d$accepted_traj[mask]) + ggtitle("SD = 1")
mask <- d$proposal_sd=="Proposal SD = 10"
plot3 <- ggAcf(x = d$accepted_traj[mask]) + ggtitle("SD = 10")

library(patchwork)
(plot1 + plot2 + plot3)
```

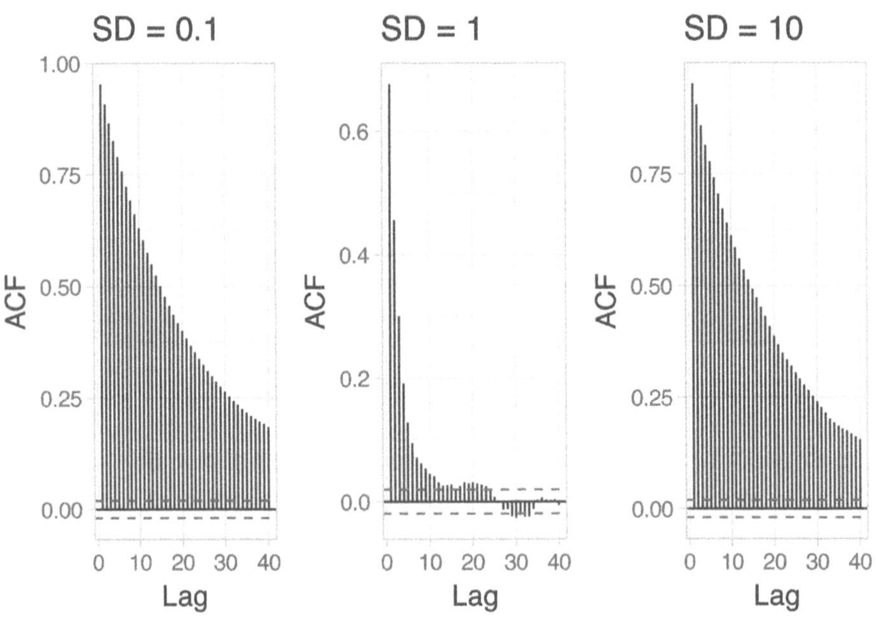

In the left and right panels, autocorrelation is strong, decreases slowly with increasing lag and mixing is bad. In the center panel, autocorrelation is weak, decreases rapidly with increasing lag and mixing is good.

Autocorrelation is not necessarily a big issue. Strongly correlated observations just require large sample sizes and therefore longer simulations. But how many iterations exactly? The effective sample size (n.eff) measures chain length while taking into account chain autocorrelation. You should check the n.eff of every parameter of interest, and of any interesting parameter combinations. In general, we need n.eff ≥ 400 independent steps to get reasonable Monte Carlo estimates of model parameters. In the animal survival example, n.eff can be calculated with the R coda::effectiveSize() function:

```r
mask <- d$proposal_sd=="Proposal SD = 0.1"
neff1 <- coda::effectiveSize(d$accepted_traj[mask])
mask <- d$proposal_sd=="Proposal SD = 1"
neff2 <- coda::effectiveSize(d$accepted_traj[mask])
mask <- d$proposal_sd=="Proposal SD = 10"
neff3 <- coda::effectiveSize(d$accepted_traj[mask])
df <- tibble("Proposal SD" = c(0.1, 1, 10),
             "n.eff" = round(c(neff1, neff2, neff3)))
df
## # A tibble: 3 x 2
##    `Proposal SD` n.eff
##            <dbl> <dbl>
## 1            0.1   224
## 2            1    1934
## 3           10     230
```

As expected, n.eff is less than the number of MCMC iterations because of autocorrelation. Only when the standard deviation of the proposal distribution is 1 is the mixing good and we get a satisfying effective sample size.

1.6.3   What if you have issues of convergence?

When diagnosing MCMC convergence, you will (very) often run into troubles. In this section, you will find some helpful tips I hope.

When mixing is bad and effective sample size is small, you may just need to increase burn-in and/or sample more. Using more informative priors might also make Markov chains converge faster by helping your MCMC sampler (e.g. the Metropolis algorithm) navigating more efficiently the parameter space. In the same spirit, picking better initial values for starting the chain does not harm. For doing that, a strategy consists in using estimates from a simpler model for which your MCMC chains do converge.

If convergence issues persist, often there is a problem with your model (also known as *the folk theorem of statistical computing* as stated by Andrew Gelman, see https://statmodeling.stat.columbia.edu/2008/05/13/the_folk_theore/). A bug in the code? A typo somewhere? A mistake in your maths? As often when coding is involved, the issue can be identified by removing complexities, and start with a simpler model until you find what the problem is.

A general advice is to see your model as a data generating tool in the first place, simulate data from it using some realistic values for the parameters, and try to recover these parameter values by fitting the model to the simulated data. Simulating from a model will help you understanding how it works, what it does not do, and the data you need to get reasonable parameter estimates (e.g. Chapter 3 and Section 7.4).

1.7 Summary

- With the Bayes' theorem, you update your beliefs (prior) with new data (likelihood) to get posterior beliefs (posterior): posterior $\propto$ likelihood $\times$ prior.

- The idea of Markov chain Monte Carlo (MCMC) is to simulate values from a Markov chain which has a stationary distribution equal to the posterior distribution you're after.

- In practice, you run a Markov chain multiple times starting from over-dispersed initial values.

- You discard iterations in an initial burn-in phase and achieve convergence when all chains reach the same regime.

- From there, you run the chains long enough and proceed with calculating Monte Carlo estimates of numerical summaries (e.g. posterior means and credible intervals) for parameters.

1.8 Suggested reading

- McCarthy [2007] and Kéry and Schaub [2011] are excellent introductions to Bayesian statistics for ecologists. The forthcoming second edition of the latter will also include NIMBLE code. See also Kéry and Kellner [2024].

- For deeper insights, I recommend Gelman and Hill [2006] which analyze data using the frequentist and Bayesian approaches side-by-side. The book by McElreath [2020] is also an excellent read. The presentation of the Metropolis algorithm in Section 1.5.3 was inspired by Albert and Hu [2019]. If you'd like to know more about Monte Carlo methods, the book Robert and Casella [2004] is a must [see also its R counterpart Robert and Casella, 2010]. See also Vehtari et al. [2021] for a discussion on $\hat{R}$ and the effective sample size.

- I also recommend Gelman et al. [2020] in which the authors offer a workflow for Bayesian analyses. They discuss model building, model comparison, model checking, model validation, model understanding and troubleshooting of computational problems.

2

NIMBLE tutorial

2.1 Introduction

In this second chapter, you will get familiar with NIMBLE, an R package that implements up-to-date MCMC algorithms for fitting complex models. NIMBLE spares you from coding the MCMC algorithms by hand, and requires only the specification of a likelihood and priors for model parameters. Should you wish to dive deeper into the mechanics, NIMBLE also got you covered and allows you to write samples, use custom functions, etc. We will illustrate NIMBLE's main features with a simple example, but the ideas hold for more complex problems.

2.2 What is NIMBLE?

NIMBLE stands for **N**umerical **I**nference for statistical **M**odels using **B**ayesian and **L**ikelihood **E**stimation (Figure 2.1). Briefly speaking, NIMBLE is an R package that implements for you MCMC algorithms to generate samples from the posterior distribution of model parameters. Freed from the burden of coding your own MCMC algorithms, you only have to specify a likelihood and priors to apply the Bayes theorem. To do so, NIMBLE makes this easy by using a syntax very similar to the R syntax, which should make your life easier. It is also a direct extension

DOI: 10.1201/9781003244097-2

FIGURE 2.1: Logo of the NIMBLE R package designed by Luke Larson.

of the BUGS language is also used by other programs like WinBUGS, OpenBUGS, and JAGS.

So why use NIMBLE you may ask? The short answer is that NIMBLE is capable of so much more than just running MCMC algorithms! First, you will work from within R, but in the background NIMBLE will translate your code in C++ for (in general) faster computation. Second, NIMBLE extends the BUGS language for writing new functions and distributions of your own, or borrow those written by others. Third, NIMBLE gives you full control of the MCMC samplers, and you may pick other algorithms than the defaults. Fourth, NIMBLE comes with a library of numerical methods other than MCMC algorithms, including sequential Monte Carlo (for particle filtering), Monte Carlo Expectation Maximization (for maximum likelihood), Hamiltonian Monte Carlo (like in program Stan), and Laplace approximation (like in program TMB). Last but not least, the development team is friendly and helpful, and based on users' feedbacks, NIMBLE folks work constantly at improving the package capabilities. The NIMBLE users google group is an open and inclusive space where everyone can receive help from the community: https://groups.google.com/g/nimble-users.

2.3   Getting started

> To run NIMBLE, you will need to:
> 1. Build a model consisting of a likelihood and priors.
> 2. Read in some data.
> 3. Specify parameters you want to make inference about.
> 4. Pick initial values for parameters to be estimated (for each chain).
> 5. Provide MCMC details namely the number of chains, the length of the burn-in period and the number of iterations following burn-in.

First things first, let's not forget to load the `nimble` package:

```r
library(nimble)
```

Note that before you can install `nimble` like any other R package, Windows users will need to install `Rtools`, and Mac users will need to install `Xcode`. More info and help trouble-shooting installation issues can be found here: https://r-nimble.org/download.

Now let's go back to our example on animal survival from the previous chapter. First step is to build our model by specifying the binomial likelihood and a uniform prior on survival probability `theta`. We use the `nimbleCode()` function and wrap code within curly brackets:

```r
model <- nimbleCode({
  # likelihood
  survived ~ dbinom(theta, released)
  # prior
  theta ~ dunif(0, 1)
  # derived quantity
  lifespan <- -1/log(theta)
})
```

You can check that the `model` R object contains your code:

```
model
## {
##     survived ~ dbinom(theta, released)
##     theta ~ dunif(0, 1)
##     lifespan <- -1/log(theta)
## }
```

In the code above, `survived` and `released` are known, only `theta` needs to be estimated. The line `survived ~ dbinom(theta, released)` states that the number of successes or animals that have survived over winter, `survived`, is distributed as (that's the ~) a binomial with `released` trials and probability of success or survival `theta`. Then the line `theta ~ dunif(0, 1)` assigns a uniform distribution between 0 and 1 as a prior to the survival probability. This is all you need, a likelihood and priors for model parameters, NIMBLE knows the Bayes theorem. The last line `lifespan <- - 1/log(theta)` calculates a quantity derived from `theta`, which is the expected lifespan assuming constant survival. If you'd like to know more about the calculation of life expectancy, check out Cook et al. [1967].

A few comments:

- The most common distributions are readily available in NIMBLE. Among others, we will use later in the book `dbeta`, `dmultinom` and `dnorm`. If you cannot find what you need in NIMBLE, you can write your own distributions as illustrated in Section 2.4.

- It does not matter in what order you write each line of code, NIMBLE uses what is called a declarative language for building models. In brief, you write code that tells NIMBLE what you want to achieve, and not how to get there. In contrast, an imperative language requires that you write what you want your program to do step by step.

- You can think of models in NIMBLE as graphs. A graph is made of relations (or edges) that can be of two types. A stochastic relation is signaled by a ~ sign and defines a random variable in the model, such as `survived` or `theta`. A deterministic relation is signaled by a <-

sign, like `lifespan`. Relations define nodes on the left – the children – in terms of other nodes on the right – the parents – and relations are directed arrows from parents to children. Such graphs are called directed acyclic graph or DAG.

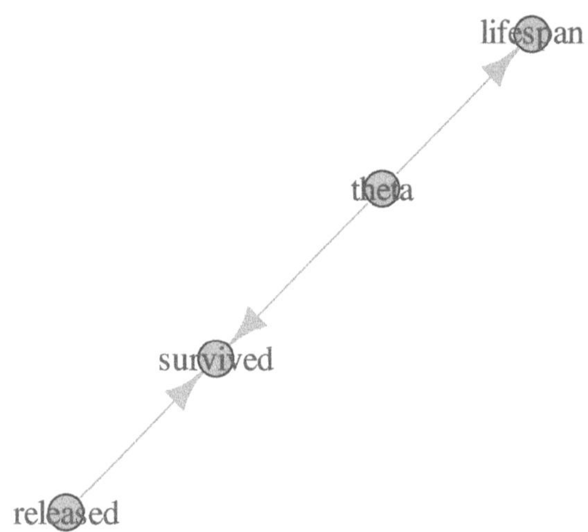

Second step in our workflow is to read in some data. We use a list in which each component corresponds to a known quantity in the model:

```r
my.data <- list(released = 57, survived = 19)
```

You can proceed with data passed this way, but you should know a little more about how NIMBLE sees data. NIMBLE distinguishes data and constants. Constants are values that do not change, e.g. vectors of known index values or the indices used to define for loops. Data are values that you might want to change, basically anything that only appears on the left of a ~. Declaring relevant values as constants is better for computational efficiency, but it is easy to forget, and fortunately NIMBLE will by itself distinguish data and constants. It will suggest you to move some data into constants to improve efficiency. I will not use the distinction between data and constants in this chapter, but in the next chapters it will become important.

Third step is to tell NIMBLE which nodes in your model you would like to keep track of, in other words the quantities you'd like to do inference about. In our model we want survival `theta` and `lifespan`:

```r
parameters.to.save <- c("theta", "lifespan")
```

In general you have many quantities in your model, including some of
little interest that are not worth monitoring, and having full control on
verbosity will prove handy.

Fourth step is to specify initial values for all model parameters. As a
bare minimum, you need initial values for all nodes that only appear
on the left side of a ~ in your code and are not given as data. To make
sure that the MCMC algorithm explores the posterior distribution, we
start different chains with different parameter values. You can specify
initial values for each chain (here we specify for three chains) in a list
and put them in yet another list:

```r
init1 <- list(theta = 0.1)
init2 <- list(theta = 0.5)
init3 <- list(theta = 0.9)
initial.values <- list(init1, init2, init3)
initial.values
## [[1]]
## [[1]]$theta
## [1] 0.1
##
##
## [[2]]
## [[2]]$theta
## [1] 0.5
##
##
## [[3]]
## [[3]]$theta
## [1] 0.9
```

Alternatively, you can write an R function that generates random initial
values:

```
initial.values <- function() list(theta = runif(1,0,1))
initial.values()
## $theta
## [1] 0.8356
```

If you are using a function to generate random initial values, it's always a good idea to set the seed in your code before you draw the initial values. For example like this:

```
my.seed <- 666
set.seed(my.seed)
```

Setting the seed makes your code reproducible, which really helps if you need to trouble-shoot it later. Initialization problems are not uncommon when working with NIMBLE, and being able to reproduce the same initial values again is very useful for solving them.

Fifth and last step, you need to tell NIMBLE the number of chains to run, say `n.chain`, how long the burn-in period should be, say `n.burnin`, and the number of iterations following the burn-in period to be used for posterior inference:

```
n.iter <- 5000
n.burnin <- 1000
n.chains <- 3
```

In NIMBLE, you specify the total number of iterations, say `n.iter`, so that the number of posterior samples per chain is `n.iter - n.burnin`. NIMBLE also allows discarding samples after burn-in, a procedure known as thinning. Thinning is fixed to 1 by default in NIMBLE so that all simulations are used to summarize posterior distributions. Link and Eaton [2012] offer a discussion of the pros and cons of thinning.

We now have all the ingredients to run our model, that is to sample from the posterior distribution of model parameters using MCMC simulations. This is accomplished using function `nimbleMCMC()`:

```r
mcmc.output <- nimbleMCMC(code = model,
                          data = my.data,
                          inits = initial.values,
                          monitors = parameters.to.save,
                          niter = n.iter,
                          nburnin = n.burnin,
                          nchains = n.chains)
## |-------------|-------------|-------------|-------------|
## |-------------------------------------------------------|
## |-------------|-------------|-------------|-------------|
## |-------------------------------------------------|
## |------------|-------------|----------|-------------|
## |-------------------------------------------------------|
```

NIMBLE goes through several steps that we will explain in Section 2.5. Function `nimbleMCMC()` takes other arguments that you might find useful. For example, one is `setSeed`. Just like with sampling initial values above, setting the seed within the MCMC call allows you to run the same chains (again), thus making your analyses reproducible and problems easier to debug (see Section 2.7.5). You can also get a summary of the outputs by specifying `summary = TRUE`. Conversely, if you would rather just get the MCMC samples back (in `coda` `mcmc` format) you can set `samplesAsCodaMCMC = TRUE`. Finally, you can suppress the progress bar if you find it too depressing when running long simulations with `progressBar = FALSE`. Check `?nimbleMCMC` for more details.

Now let's inspect what we have in `mcmc.output`:

```r
str(mcmc.output)
## List of 3
##  $ chain1: num [1:4000, 1:2] 0.907 0.907 0.907 0.907 0.853 ...
##   ..- attr(*, "dimnames")=List of 2
##   .. ..$ : NULL
##   .. ..$ : chr [1:2] "lifespan" "theta"
##  $ chain2: num [1:4000, 1:2] 0.787 0.894 1.291 1.388 1.388 ...
##   ..- attr(*, "dimnames")=List of 2
```

```
##    .. ..$ : NULL
##    .. ..$ : chr [1:2] "lifespan" "theta"
##   $ chain3: num [1:4000, 1:2] 0.745 0.745 0.745 0.886 1.136 ...
##    ..- attr(*, "dimnames")=List of 2
##    .. ..$ : NULL
##    .. ..$ : chr [1:2] "lifespan" "theta"
```

The R object `mcmc.output` is a list with three components, one for each MCMC chain. Let's have a look to `chain1` for example:

```
dim(mcmc.output$chain1)
## [1] 4000    2
head(mcmc.output$chain1)
##        lifespan  theta
## [1,]   0.9069 0.3320
## [2,]   0.9069 0.3320
## [3,]   0.9069 0.3320
## [4,]   0.9069 0.3320
## [5,]   0.8526 0.3095
## [6,]   0.7987 0.2859
```

Each component of the list is a matrix. In rows, you have 4000 samples from the posterior distribution of `theta`, which corresponds to `n.iter` – `n.burnin` iterations. In columns, you have the quantities we monitor, `theta` and `lifespan`. From there, you can compute the posterior mean of `theta`:

```
mean(mcmc.output$chain1[,'theta'])
## [1] 0.3407
```

You can also obtain the 95% credible interval for `theta`:

```
quantile(mcmc.output$chain1[,'theta'], probs = c(2.5, 97.5)/100)
##    2.5%  97.5%
## 0.2219 0.4620
```

Let's visualize the posterior distribution of `theta` with a histogram:

```r
mcmc.output$chain1[,"theta"] %>%
  as_tibble() %>%
  ggplot() +
  geom_histogram(aes(x = value), color = "white") +
  labs(x = "survival probability")
```

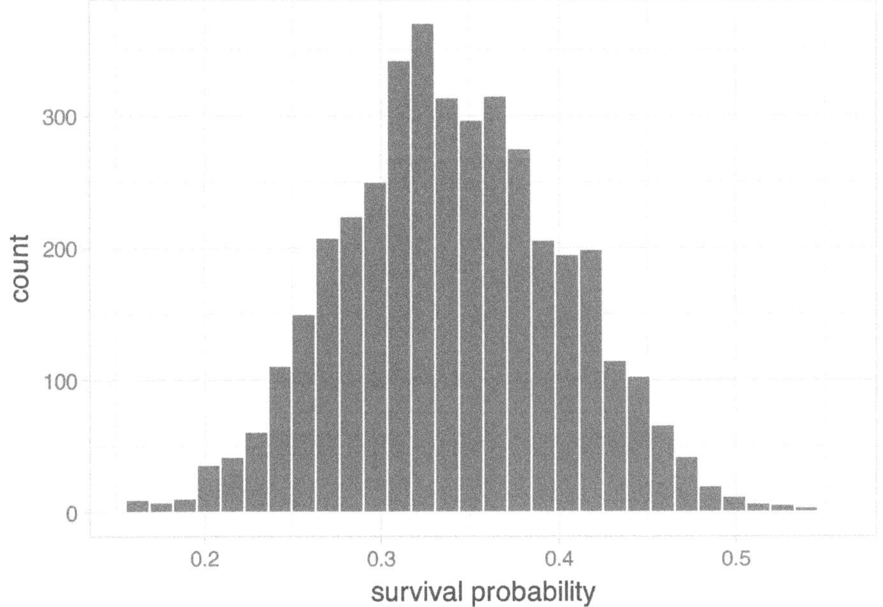

There are less painful ways of doing posterior inference. In this book, I will use the R package `MCMCvis` to summarize and visualize MCMC outputs, but there are other perfectly valid options out there like `ggmcmc`, `bayesplot` and `basicMCMCplots`.

Let's load the package `MCMCvis`:

```r
library(MCMCvis)
```

To get the most common numerical summaries, the function `MCMCsummary()` does the job:

```
MCMCsummary(object = mcmc.output, round = 2)
##             mean   sd 2.5%  50% 97.5% Rhat n.eff
## lifespan 0.94 0.17 0.66 0.92  1.32    1  2513
## theta    0.34 0.06 0.22 0.34  0.47    1  2533
```

You can use a caterpillar plot to visualize the posterior distributions of
theta with `MCMCplot()`:

```
MCMCplot(object = mcmc.output,
         params = 'theta')
```

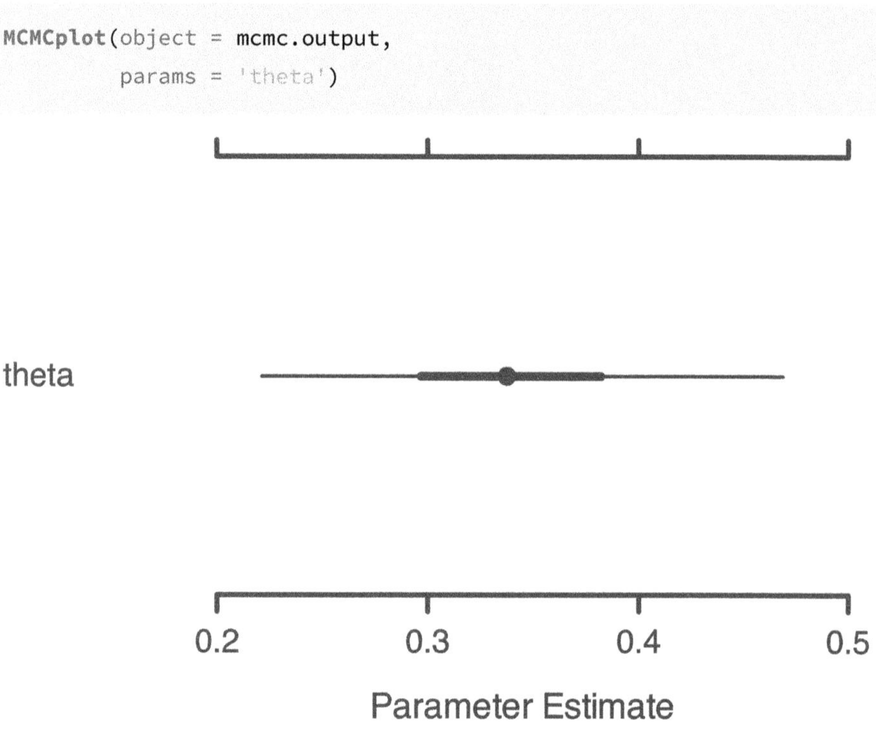

The point represents the posterior median, the thick line is the 50%
credible interval and the thin line the 95% credible interval.

Visualization of a MCMC chain itself, i.e. the values of posterior samples plotted against iteration number, is called a trace. The trace and
posterior density of theta can be obtained with `MCMCtrace()`:

```
MCMCtrace(object = mcmc.output,
          pdf = FALSE, # no export to PDF
```

```
ind = TRUE, # separate density lines per chain
params = "theta")
```

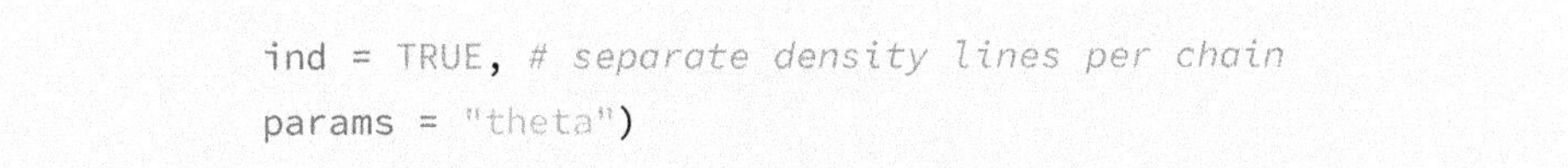

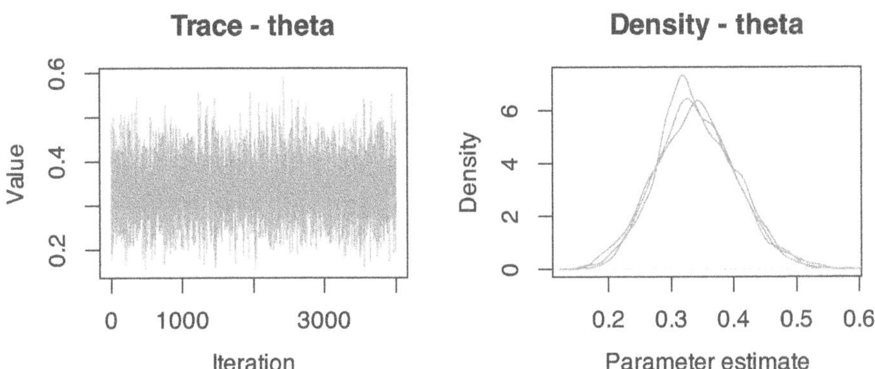

We use the trace and density plots for assessing convergence and get an idea of whether there may be any estimation issues (see Section 1.6).

You can also add the diagnostics of convergence we discussed in the previous chapter:

```
MCMCtrace(object = mcmc.output,
          pdf = FALSE,
          ind = TRUE,
          Rhat = TRUE, # add Rhat
          n.eff = TRUE, # add eff sample size
          params = "theta")
```

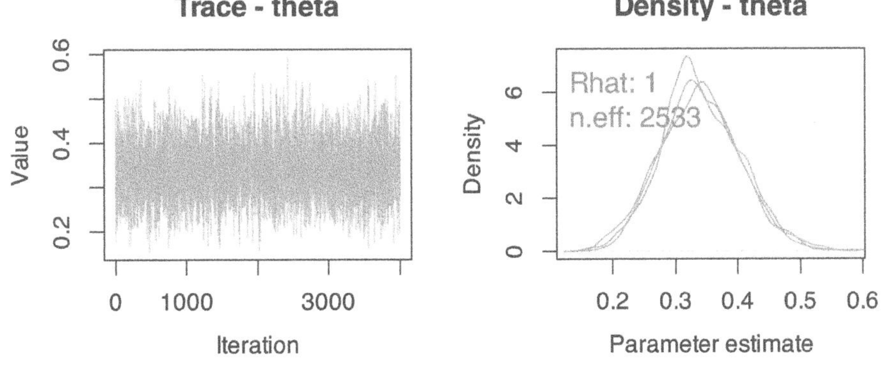

We calculated lifespan directly in our model with `lifespan <- -1/log(theta)`. But you can also calculate this quantity from outside NIMBLE. This is a nice by-product of using MCMC simulations: You can obtain the posterior distribution of any quantity that is a function of your model parameters by applying this function to samples from the posterior distribution of these parameters. Especially when working with big models/data, it is recommended to keep any calculations that can be made "post-hoc" using the posterior samples outside of NIMBLE as this lessens memory load. In our example, all you need is samples from the posterior distribution of `theta`, which we pool between the three chains with:

```r
theta_samples <- c(mcmc.output$chain1[,'theta'],
                   mcmc.output$chain2[,'theta'],
                   mcmc.output$chain3[,'theta'])
```

To get samples from the posterior distribution of lifespan, we apply the function to calculate lifespan to the samples from the posterior distribution of survival:

```r
lifespan <- -1/log(theta_samples)
```

As usual then, you can calculate the posterior mean and 95% credible interval:

```r
mean(lifespan)
## [1] 0.9398
quantile(lifespan, probs = c(2.5, 97.5)/100)
##    2.5%   97.5%
## 0.6629 1.3194
```

You can also visualize the posterior distribution of lifespan:

```r
lifespan %>%
  as_tibble() %>%
  ggplot() +
  geom_histogram(aes(x = value), color = "white") +
  labs(x = "lifespan")
```

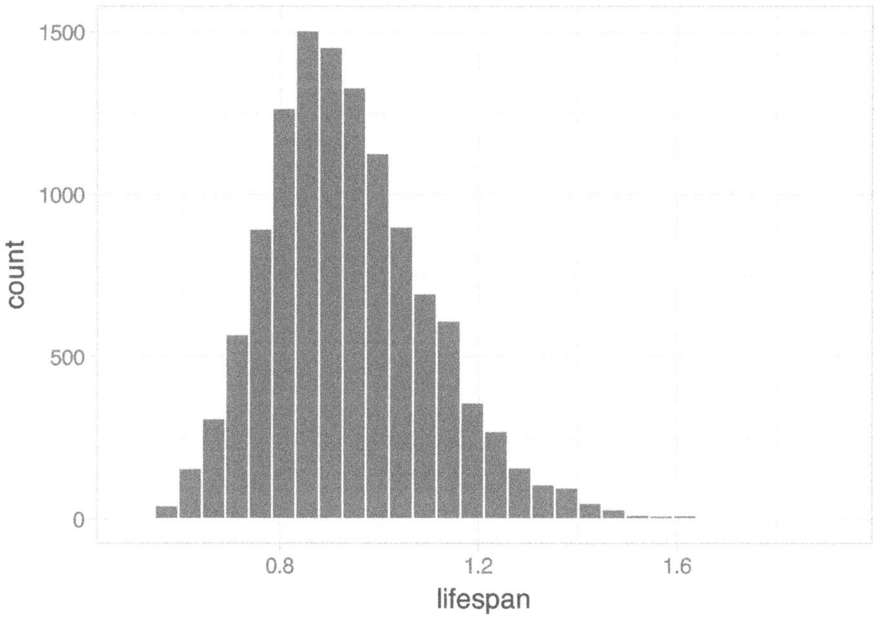

Now you're good to go. For convenience I have summarized the steps above in the box below. The NIMBLE workflow provided with `nimbleM-CMC()` allows you to build models and make inference. This is what you can achieve with other software like WinBUGS or JAGS.

NIMBLE workflow:

```r
# model building
model <- nimbleCode({
  # likelihood
  survived ~ dbinom(theta, released)
  # prior
  theta ~ dunif(0, 1)
  # derived quantity
  lifespan <- -1/log(theta)
})
# read in data
my.data <- list(released = 57, survived = 19)
# specify parameters to monitor
parameters.to.save <- c("theta", "lifespan")
# pick initial values
initial.values <- function() list(theta = runif(1,0,1))
# specify MCMC details
n.iter <- 5000
n.burnin <- 1000
n.chains <- 3
# run NIMBLE
mcmc.output <- nimbleMCMC(code = model,
                          data = my.data,
                          inits = initial.values,
                          monitors = parameters.to.save,
                          niter = n.iter,
                          nburnin = n.burnin,
                          nchains = n.chains)
# calculate numerical summaries
MCMCsummary(object = mcmc.output, round = 2)
# visualize parameter posterior distribution
MCMCplot(object = mcmc.output,
         params = 'theta')
# check convergence
MCMCtrace(object = mcmc.output,
          pdf = FALSE, # no export to PDF
          ind = TRUE, # separate density lines per chain
          params = "theta")
```

But NIMBLE is more than just another MCMC engine. It provides a programming environment so that you have full control when building models and estimating parameters. NIMBLE allows you to write your own functions and distributions to build models, and to choose alternative MCMC samplers or code new ones. This flexibility often comes with faster convergence and often faster runtime.

I have to be honest, learning these improvements over other software takes some reading and experimentation, and it might well be that you do not need to use any of these features. And it's fine. In the next sections, I cover some of this advanced material. You may skip these sections and go back to this material later if you need it.

2.4   Programming

In NIMBLE you can write and use your own functions, or use existing R or C/C++ functions. This allows you to customize models the way you want.

2.4.1   NIMBLE functions

NIMBLE provides `nimbleFunctions` for programming. A `nimbleFunction` is like an R function, plus it can be compiled for faster computation. Going back to our animal survival example, we can write a `nimbleFunction` to compute lifespan:

```
computeLifespan <- nimbleFunction(
    run = function(theta = double(0)) { # type declarations
        ans <- -1/log(theta)
        return(ans)
        returnType(double(0))   # return type declaration
    } )
```

Within the nimbleFunction, the `run` section gives the function to be executed. It is written in the NIMBLE language. The `theta = double(0)` and `returnType(double(0))` arguments tell NIMBLE that the input and output are single numeric values (scalars). Alternatively, double(1) and double(2) are for vectors and matrices, while logical(), integer() and character() are for logical, integer and character values.

You can use your nimbleFunction in R:

```
computeLifespan(0.8)
## [1] 4.481
```

You can compile it and use the C++ code for faster computation:

```
CcomputeLifespan <- compileNimble(computeLifespan)
CcomputeLifespan(0.8)
## [1] 4.481
```

You can also use your nimbleFunction in a model:

```
model <- nimbleCode({
  # likelihood
  survived ~ dbinom(theta, released)
  # prior
  theta ~ dunif(0, 1)
  # derived quantity
  lifespan <- computeLifespan(theta)
})
```

The rest of the workflow remains the same:

```
my.data <- list(survived = 19, released = 57)
parameters.to.save <- c("theta", "lifespan")
initial.values <- function() list(theta = runif(1,0,1))
n.iter <- 5000
n.burnin <- 1000
n.chains <- 3
mcmc.output <- nimbleMCMC(code = model,
                          data = my.data,
```

```
                    inits = initial.values,
                    monitors = parameters.to.save,
                    niter = n.iter,
                    nburnin = n.burnin,
                    nchains = n.chains)
## |------------------|-----------------|-----------------|-----------------|
## |------------------|-----------------|-----------------|-----------------|
## |------------------|-----------------|-----------------|-----------------|
## |------------------|-----------------|-----------------|-----------------|
## |------------------|-----------------|-----------------|-----------------|
## |------------------|-----------------|-----------------|-----------------|
MCMCsummary(object = mcmc.output, round = 2)
##          mean   sd 2.5%  50% 97.5% Rhat n.eff
## lifespan 0.94 0.16 0.66 0.92  1.31    1  2593
## theta    0.34 0.06 0.22 0.34  0.47    1  2652
```

With `nimbleFunctions`, you can mimic basic R syntax, do linear algebra
(e.g. compute eigenvalues), operate on vectors and matrices (e.g. inverse a matrix), use logical operators (e.g. and/or) and flow control
(e.g. if-else). There is also a long list of common and less common distributions that can be used with `nimbleFunctions`.

To learn everything you need to know on writing `nimbleFunctions`, make
sure to read chapter 11 of the NIMBLE manual at https://r-nimble.org
/manual/cha-RCfunctions.html.

2.4.2 Calling R/C++ functions

If you're like me, and too lazy to write your own functions, you can rely
on the scientific community and use existing C, C++ or R code. The
trick is to write a `nimbleFunction` that wraps access to that code which
can then be used by NIMBLE. As an example, imagine you'd like to use
an R function `myfunction()`, either a function you wrote yourself, or a
function available in your favorite R package:

```r
myfunction <- function(x) {
  -1/log(x)
}
```

Now wrap this function using `nimbleRcall()` or `nimbleExternalCall()` for a C or C++ function:

```r
Rmyfunction <- nimbleRcall(prototype = function(x = double(0)){},
                           Rfun = 'myfunction',
                           returnType = double(0))
```

In the call to `nimbleRcall()` above, the argument `prototype` specifies inputs (a single numeric value `double(0)`) of the R function `Rfun` that generates outputs `returnType` (a single numeric value `double(0)`).

Now you can call your R function from a model (or any `nimbleFunctions`):

```r
model <- nimbleCode({
  # likelihood
  survived ~ dbinom(theta, released)
  # prior
  theta ~ dunif(0, 1)
  lifespan <- Rmyfunction(theta)
})
```

The rest of the workflow remains the same:

```r
my.data <- list(survived = 19, released = 57)
parameters.to.save <- c("theta", "lifespan")
initial.values <- function() list(theta = runif(1,0,1))
n.iter <- 5000
n.burnin <- 1000
n.chains <- 3
mcmc.output <- nimbleMCMC(code = model,
                          data = my.data,
```

```
                              inits = initial.values,
                              monitors = parameters.to.save,
                              niter = n.iter,
                              nburnin = n.burnin,
                              nchains = n.chains)
## |----------------|----------------|---------------|---------------|
## |--------------------------------------------------------------------|
## |----------------|----------------|---------------|---------------|
## |--------------------------------------------------------------|
## |----------------|---------------|---------------|-------------|
## |--------------------------------------------------------------|
MCMCsummary(object = mcmc.output, round = 2)
##            mean    sd 2.5%  50% 97.5% Rhat n.eff
## lifespan 0.94 0.16 0.68 0.92  1.29     1  2597
## theta    0.34 0.06 0.23 0.34  0.46     1  2643
```

Evaluating an R function from within NIMBLE slows down MCMC
sampling, but if you can live with it, the cost is easily offset by the
convenience of being able to use existing R functions.

2.4.3   User-defined distributions

With `nimbleFunctions` you can provide user-defined distributions to
NIMBLE. You need to write functions for density (d) and simulation
(r) for your distribution. As an example, we write our own binomial
distribution:

```
# density
dmybinom <- nimbleFunction(
  run = function(x = double(0),
                 size = double(0),
                 prob = double(0),
                 log = integer(0, default = 1)) {
    returnType(double(0))
    # compute binomial coefficient = size!/[x! (n-x)!] and take log
    lchoose <- lfactorial(size)-lfactorial(x)-lfactorial(size-x)
```

```r
    # binomial density function =
    # size! / [x! (n-x)!] * prob^x * (1-prob)^(size-x) and take log
    logProb <- lchoose + x * log(prob) + (size - x) * log(1 - prob)
    if(log) return(logProb)
    else return(exp(logProb))
  })
# simulation using the coin flip method (p. 524 in Devroye 1986)
# note: the n argument is required by NIMBLE
# but is not used, default is 1
rmybinom <- nimbleFunction(
  run = function(n = integer(0, default = 1),
                 size = double(0),
                 prob = double(0)) {
    x <- 0
    y <- runif(n = size, min = 0, max = 1)
    for (j in 1:size){
      if (y[j] < prob){
        x <- x + 1
      }else{
        x <- x
      }
    }
    returnType(double(0))
    return(x)
  })
```

You need to define the `nimbleFunctions` in R's global environment for
them to be accessed:

```r
assign('dmybinom', dmybinom, .GlobalEnv)
assign('rmybinom', rmybinom, .GlobalEnv)
```

You can try out your function and simulate a single random value (n = 1
by default) from a binomial distribution with size 5 and probability 0.1:

```
rmybinom(size = 5, prob = 0.1)
## [1] 0
```

All set. You can run your workflow:

```
model <- nimbleCode({
  # likelihood
  survived ~ dmybinom(prob = theta, size = released)
  # prior
  theta ~ dunif(0, 1)
})
my.data <- list(released = 57, survived = 19)
initial.values <- function() list(theta = runif(1,0,1))
n.iter <- 5000
n.burnin <- 1000
n.chains <- 3
mcmc.output <- nimbleMCMC(code = model,
                          data = my.data,
                          inits = initial.values,
                          niter = n.iter,
                          nburnin = n.burnin,
                          nchains = n.chains)
## |-------------|-------------|-------------|-------------|
## |-------------------------------------------------------|
## |-------------|-------------|-------------|-------------|
## |-------------------------------------------------------|
## |-------------|-------------|-------------|-------------|
## |-------------------------------------------------------|
MCMCsummary(mcmc.output)
##        mean      sd    2.5%     50%   97.5% Rhat n.eff
## theta 0.34 0.05976  0.2286  0.3378  0.4598    1  2970
```

Having `nimbleFunctions` offers infinite possibilities to customize your models and algorithms. Besides what we covered already, you can write your own samplers. We will see an example in a minute, but I first need to tell you more about the NIMBLE workflow.

2.5 Under the hood

So far, you have used `nimbleMCMC()` which runs the default MCMC workflow. This is perfectly fine for most applications. However, in some situations you need to customize the MCMC samplers to improve or speed up convergence. NIMBLE allows you to look under the hood by using a detailed workflow in several steps: `nimbleModel()`, `configureMCMC()`, `buildMCMC()`, `compileNimble()` and `runMCMC()`. Note that `nimbleMCMC()` does all of this at once.

We write the model code, read in data and pick initial values as before:

```r
model <- nimbleCode({
  # likelihood
  survived ~ dbinom(theta, released)
  # prior
  theta ~ dunif(0, 1)
  # derived quantity
  lifespan <- -1/log(theta)
})
my.data <- list(survived = 19, released = 57)
initial.values <- list(theta = 0.5)
```

First step is to create the model as an R object (uncompiled model) with `nimbleModel()`:

```r
survival <- nimbleModel(code = model,
                        data = my.data,
                        inits = initial.values)
```

You can look at its nodes:

```r
survival$getNodeNames()
## [1] "theta"    "lifespan" "survived"
```

You can look at the values stored at each node:

```r
survival$theta
## [1] 0.5
survival$survived
## [1] 19
survival$lifespan
## [1] 1.443
# this is -1/log(0.5)
```

We can also calculate the log-likelihood at the initial value for `theta`:

```r
survival$calculate()
## [1] -5.422
# this is dbinom(x = 19, size = 57, prob = 0.5, log = TRUE)
```

The ability in NIMBLE to access the nodes of your model and to evaluate the model likelihood can help you in identifying bugs in your code. For example, if we provide a negative initial value for `theta`, `survival$calculate()` returns NA:

```r
survival <- nimbleModel(code = model,
                        data = my.data,
                        inits = list(theta = -0.5))
survival$calculate()
## [1] NaN
```

As another example, if we convey in the data the information that more animals survived than were released, we'll get an infinity value for the log-likelihood:

```r
my.data <- list(survived = 61, released = 57)
initial.values <- list(theta = 0.5)
survival <- nimbleModel(code = model,
                        data = my.data,
                        inits = initial.values)
```

```
survival$calculate()
## [1] -Inf
```

As a check that the model is correctly initialized and that your code is without bugs, the call to `model$calculate()` should return a number and not NA or -Inf:

```
my.data <- list(survived = 19, released = 57)
initial.values <- list(theta = 0.5)
survival <- nimbleModel(code = model,
                        data = my.data,
                        inits = initial.values)
survival$calculate()
## [1] -5.422
```

You can obtain the graph of the model with:

```
survival$plotGraph()
```

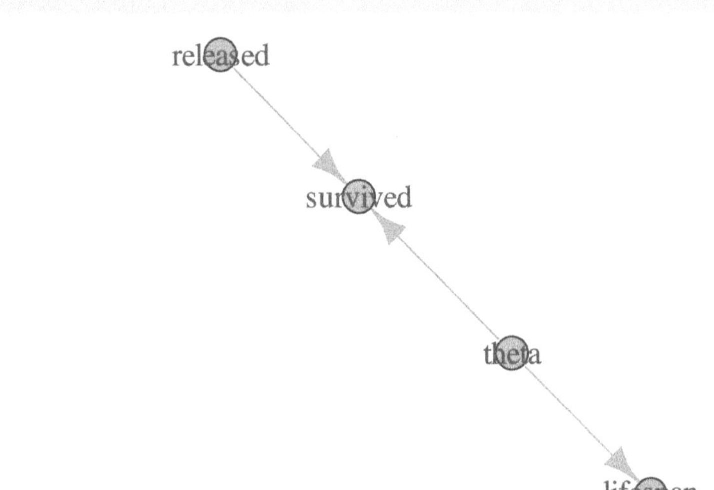

Second we compile the model with `compileNimble()`:

```
Csurvival <- compileNimble(survival)
```

With `compileNimble()`, the C++ code is generated, compiled and loaded back into R so that it can be used in R (compiled model):

```
Csurvival$theta
## [1] 0.5
```

Now you have two versions of the model, `survival` is in R and `Csurvival` in C++. Being able to separate the steps of model building and parameter estimation is a strength of NIMBLE. This gives you a lot of flexibility at both steps. For example, imagine you would like to fit your model with maximum likelihood, then you can do it by wrapping your model in an R function that gets the likelihood and maximize this function. Using the C version of the model, you can write:

```
# function for negative log-likelihood to minimize
f <- function(par) {
    Csurvival[['theta']] <- par # assign par to theta
    ll <- Csurvival$calculate() # update log-likelihood w/ par value
    return(-ll) # return negative log-likelihood
}
# evaluate function at 0.5 and 0.9
f(0.5)
## [1] 5.422
f(0.9)
## [1] 55.41
# minimize function
out <- optimize(f, interval = c(0,1))
round(out$minimum, 2)
## [1] 0.33
```

By maximizing the likelihood (or minimizing the negative log-likelihood), you obtain the maximum likelihood estimate of animal survival, which is exactly 19 surviving animals over 57 released animals or 0.33.

Third we create a MCMC configuration for our model with `configureM-CMC()`:

```
survivalConf <- configureMCMC(survival)
## ===== Monitors =====
## thin = 1: theta
## ===== Samplers =====
## RW sampler (1)
##    - theta
```

This steps tells you the nodes that are monitored by default, and the MCMC samplers than have been assigned to them. Here `theta` is monitored, and samples from its posterior distribution are simulated with a random walk sampler similar to the Metropolis sampler we coded in the previous chapter in Section 1.5.3.

To monitor `lifespan` in addition to `theta`, you write:

```
survivalConf$addMonitors(c("lifespan"))
## thin = 1: lifespan, theta
survivalConf
## ===== Monitors =====
## thin = 1: lifespan, theta
## ===== Samplers =====
## RW sampler (1)
##    - theta
```

Third, we create a MCMC function with `buildMCMC()` and compile it with `compileNimble()`:

```
survivalMCMC <- buildMCMC(survivalConf)
CsurvivalMCMC <- compileNimble(survivalMCMC, project = survival)
```

Note that models and `nimbleFunctions` need to be compiled before they can be used to specify a project.

Fourth, we run NIMBLE with `runMCMC()`:

```
n.iter <- 5000
n.burnin <- 1000
samples <- runMCMC(mcmc = CsurvivalMCMC,
                   niter = n.iter,
                   nburnin = n.burnin)
## |---------------|---------------|---------------|---------------|
## |---------------------------------------------------------------|
```

We run a single chain but `runMCMC()` allows you to use multiple chains as with `nimbleMCMC()`.

You can look into `samples` which contains values simulated from the posterior distribution of the parameters we monitor:

```
head(samples)
##        lifespan  theta
## [1,]    0.9093 0.3330
## [2,]    0.9093 0.3330
## [3,]    0.9093 0.3330
## [4,]    1.2095 0.4374
## [5,]    1.2095 0.4374
## [6,]    1.1835 0.4296
```

From here, you can obtain numerical summaries with `samplesSummary()` (or `MCMCvis::MCMCsummary()`):

```
samplesSummary(samples)
##                Mean Median St.Dev. 95%CI_low 95%CI_upp
## lifespan 0.9357 0.9194 0.16117    0.6831    1.2969
## theta    0.3386 0.3370 0.06128    0.2313    0.4625
```

I have summarized the steps above in the box below.

Detailed NIMBLE workflow:

```r
# model building
model <- nimbleCode({
  survived ~ dbinom(theta, released) # likelihood
  theta ~ dunif(0, 1) # prior
  lifespan <- -1/log(theta) # derived quantity
})
# read in data
my.data <- list(released = 57, survived = 19)
# pick initial values
initial.values <- function() list(theta = runif(1,0,1))
# create model as an R object (uncompiled model)
survival <- nimbleModel(code = model,
                        data = my.data,
                        inits = initial.values())
# compile model
Csurvival <- compileNimble(survival)
# create a MCMC configuration
survivalConf <- configureMCMC(survival)
# add lifespan to list of parameters to monitor
survivalConf$addMonitors(c("lifespan"))
# create a MCMC function and compile it
survivalMCMC <- buildMCMC(survivalConf)
CsurvivalMCMC <- compileNimble(survivalMCMC, project=survival)
# specify MCMC details
n.iter <- 5000; n.burnin <- 1000; n.chains <- 2
# run NIMBLE
samples <- runMCMC(mcmc = CsurvivalMCMC,
                   niter = n.iter,
                   nburnin = n.burnin,
                   nchain = n.chains)
# calculate numerical summaries
MCMCsummary(object = samples, round = 2)
# visualize parameter posterior distribution
MCMCplot(object = samples, params = 'theta')
# check convergence
MCMCtrace(object = samples, params = "theta",
          pdf = FALSE, # no export to PDF
          ind = TRUE) # separate density lines per chain
```

At first glance, using several steps instead of doing all these at once with `nimbleMCMC()` seems odds. Why is it useful? Mastering the whole sequence of steps allows you to play around with samplers, by changing the samplers NIMBLE picks by default, or even writing your own samplers.

2.6 MCMC samplers

2.6.1 Default samplers

What is the default sampler used by NIMBLE in our example? You can answer this question by inspecting the MCMC configuration obtained with `configureMCMC()`:

```r
#survivalConf <- configureMCMC(survival)
survivalConf$printSamplers()
## [1] RW sampler: theta
```

Now that we have control on the MCMC configuration, let's mess it up. We start by removing the default sampler:

```r
survivalConf$removeSamplers(c('theta'))
survivalConf$printSamplers()
```

And we change it for a slice sampler:

```r
survivalConf$addSampler(target = c('theta'),
                        type = 'slice')
survivalConf$printSamplers()
## [1] slice sampler: theta
```

Now you can resume the workflow:

```r
# create a new MCMC function and compile it:
survivalMCMC2 <- buildMCMC(survivalConf)
# to compile new functions into existing project,
# we need to reset nimbleFunctions
CsurvivalMCMC2 <- compileNimble(survivalMCMC2,
                                project = survival,
                                resetFunctions = TRUE)
# run NIMBLE:
samples2 <- runMCMC(mcmc = CsurvivalMCMC2,
                    niter = n.iter,
                    nburnin = n.burnin)
## |-------------------|-----------------|-----------------|-----------------|
## |----------------------------------------------------------------------|
# obtain numerical summaries:
samplesSummary(samples2)
##               Mean Median St.Dev. 95%CI_low 95%CI_upp
## lifespan 0.9357 0.9231 0.16002    0.6645    1.2826
## theta    0.3387 0.3385 0.06098    0.2221    0.4586
```

NIMBLE implements many samplers, and a list is available with `?samplers`. For example, high correlation in (regression) parameters can make independent samplers inefficient. In that situation, block sampling might help which consists in proposing candidate values from a multivariate distribution that acknowledges correlation between parameters.

2.6.2   User-defined samplers

Allowing you to code your own sampler is another topic on which NIMBLE thrives. As an example, we focus on the Metropolis algorithm of Section 1.5.3 which we coded in R. In this section, we make it a `nimbleFunction` so that we can use it within our model:

```r
my_metropolis <- nimbleFunction(
  name = 'my_metropolis', # fancy name for our MCMC sampler
  contains = sampler_BASE,
```

```r
setup = function(model, mvSaved, target, control) {
  # i) get dependencies for 'target' in 'model'
  calcNodes <- model$getDependencies(target)
  # ii) get sd of proposal distribution
  scale <- control$scale
},
run = function() {
  # (1) log-lik at current value
  initialLP <- model$getLogProb(calcNodes)
  # (2) current parameter value
  current <- model[[target]]
  # (3) logit transform
  lcurrent <- log(current / (1 - current))
  # (4) propose candidate value
  lproposal <- lcurrent + rnorm(1, mean = 0, scale)
  # (5) back-transform
  proposal <- plogis(lproposal)
  # (6) plug candidate value in model
  model[[target]] <<- proposal
  # (7) log-lik at candidate value
  proposalLP <- model$calculate(calcNodes)
  # (8) compute lik ratio on log scale
  lMHR <- proposalLP - initialLP
  # (9) spin continuous spinner and compare to ratio
  if(runif(1,0,1) < exp(lMHR)) {
    # (10) if candidate value is accepted, update current value
    copy(from = model, to = mvSaved, nodes = calcNodes,
         logProb = TRUE, row = 1)
  } else {
    ## (11) if candidate value is accepted, keep current value
    copy(from = mvSaved, to = model, nodes = calcNodes,
         logProb = TRUE, row = 1)
  }
},
methods = list(
  reset = function() {}
```

```
  )
)
```

Compared to `nimbleFunctions` we wrote earlier, `my_metropolis()` contains a `setup` function which i) gets the dependencies of the parameter to update in the `run` function with Metropolis, the target node, that would be `theta` in our example and ii) extracts control parameters, that would be `scale` the standard deviation of the proposal distribution in our example. Then the `run` function implements the steps of the Metropolis algorithm: (1) get the log-likelihood function evaluated at the current value, (2) get the current value, (3) apply the logit transform to it, (4) propose a candidate value by perturbing the current value with some normal noise controlled by the standard deviation `scale`, (5) back-transform the candidate value and (6) plug it in the model, (7) calculate the log-likelihood function at the candidate value, (8) compute the Metropolis ratio on the log scale, (9) compare output of a spinner and the Metropolis ratio to decide whether to (10) accept the candidate value and copy from the model to `mvSaved` or (11) reject it and keep the current value by copying from `mvSaved` to the model. Because this `nimbleFunction` is to be used as a MCMC sampler, several constraints need to be respected like having a `contains = sampler_BASE` statement or using the four arguments `model`, `mvSaved`, `target` and `control` in the `setup` function. Of course, NIMBLE implements a more advanced and efficient version of the Metropolis algorithm, you can look into it at https://github.com/cran/nimble/blob/master/R/MCMC_samplers.R#L184.

Now that we have our user-defined MCMC algorithm, we can change the default sampler for our new sampler as in Section 2.6.1. We start from scratch:

```r
model <- nimbleCode({
  # likelihood
  survived ~ dbinom(theta, released)
  # prior
  theta ~ dunif(0, 1)
})
```

```r
my.data <- list(survived = 19, released = 57)
initial.values <- function() list(theta = runif(1,0,1))
survival <- nimbleModel(code = model,
                        data = my.data,
                        inits = initial.values())
Csurvival <- compileNimble(survival)
survivalConf <- configureMCMC(survival)
## ===== Monitors =====
## thin = 1: theta
## ===== Samplers =====
## RW sampler (1)
##    - theta
```

We print the samplers used by default, remove the default sampler for
`theta`, replace it with our `my_metropolis()` sampler with the standard
deviation of the proposal distribution set to 0.1, and print again to
make sure NIMBLE now uses our new sampler:

```r
survivalConf$printSamplers()
## [1] RW sampler: theta
survivalConf$removeSamplers(c('theta'))
survivalConf$addSampler(target = 'theta',
                        type = 'my_metropolis',
                        control = list(scale = 0.1))
# scale is standard deviation of proposal distribution
survivalConf$printSamplers()
## [1] my_metropolis sampler: theta,   scale: 0.10000000000000001
```

The rest of the workflow is unchanged:

```r
survivalMCMC <- buildMCMC(survivalConf)
CsurvivalMCMC <- compileNimble(survivalMCMC,
                               project = survival)
samples <- runMCMC(mcmc = CsurvivalMCMC,
                   niter = 5000,
```

```
                         nburnin = 1000)
## |------------------|------------------|------------------|------------------|
## |-----------------------------------------------------------------------------|
samplesSummary(samples)
##           Mean  Median  St.Dev.  95%CI_low  95%CI_upp
## theta    0.339  0.3377  0.05592     0.2374     0.4528
```

You can re-run the analysis by setting the standard deviation of the proposal to different values, say 1 and 10, and compare the results to traceplots we obtained with our R implementation of the Metropolis algorithm in the previous chapter:

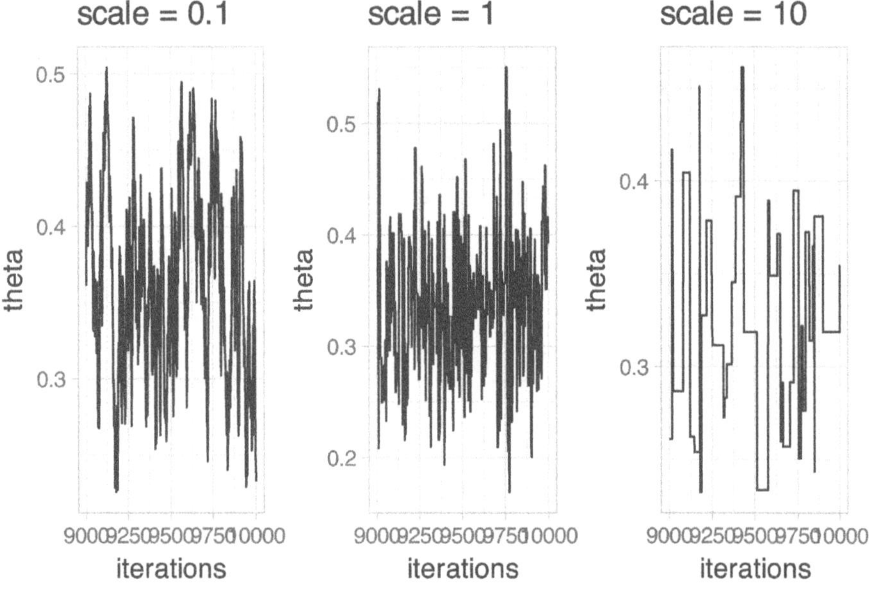

2.7   Tips and tricks

Before closing this chapter on NIMBLE, I thought it'd be useful to have a section gathering a few tips and tricks that would make your life easier.

2.7.1 Precision vs standard deviation

In other software like JAGS, the normal distribution is parameterized with mean `mu` and a parameter called precision, often denoted `tau`, the inverse of the variance you are used to. Say we use a normal prior on some parameter `epsilon` with `epsilon ~ dnorm(mu, tau)`. We'd like this prior to be vague, therefore `tau` should be small, say 0.01 so that the variance of the normal distribution is large, 1/0.01 = 100 here. This subtlety is the source of problems (and frustration) when you forget that the second parameter is precision and use `epsilon ~ dnorm(mu, 100)`, because then the variance is actually 1/100 = 0.01 and the prior is very informative, and peaked on `mu`. In NIMBLE you can use this parameterization as well as the more natural parameterization `epsilon ~ dnorm(mu, sd = 100)` which avoids confusion.

2.7.2 Indexing

NIMBLE does not guess the dimensions of objects. In other software like JAGS you can write `sum.x <- sum(x[])` to calculate the sum over all components of `x`. In NIMBLE you need to write `sum.x <- sum(x[1:n])` to sum the components of `x` from 1 up to n. Specifying dimensions can be annoying, but I find it useful as it forces me to think of what I am doing and to keep my code self-explaining.

2.7.3 Faster compilation

You might have noticed that compilation in NIMBLE takes time. When you have large models (with lots of nodes), compilation can take forever. You can set `calculate = FALSE` in `nimbleModel()` to disable the calculation of all deterministic nodes and log-likelihood. The downside of not doing the `calculate()`, is that you might not be able to identify issues that could help you save time in the long run. You can also use `useConjugacy = FALSE` in `configureMCMC()` to disable the search for conjugate samplers. With the animal survival example, you would do:

```r
model <- nimbleCode({
  # likelihood
  survived ~ dbinom(theta, released)
```

```r
  # prior
  theta ~ dunif(0, 1)
})
my.data <- list(survived = 19, released = 57)
initial.values <- function() list(theta = runif(1,0,1))
# first tip is calculate = FALSE
survival <- nimbleModel(code = model,
                        data = my.data,
                        inits = initial.values(),
                        calculate = FALSE)
Csurvival <- compileNimble(survival)
survivalConf <- configureMCMC(survival)
## ===== Monitors ======
## thin = 1: theta
## ===== Samplers ======
## RW sampler (1)
##    - theta
# second tip is useConjugacy = FALSE
survivalMCMC <- buildMCMC(survivalConf, useConjugacy = FALSE)
CsurvivalMCMC <- compileNimble(survivalMCMC,
                               project = survival)
samples <- runMCMC(mcmc = CsurvivalMCMC,
                   niter = 5000,
                   nburnin = 1000)
## |-------------|-------------|-------------|-------------|
## |-----------------------------------------------------|
samplesSummary(samples)
##         Mean Median St.Dev. 95%CI_low 95%CI_upp
## theta 0.3402 0.3391 0.06029    0.2258    0.4616
```

2.7.4   Updating MCMC chains

Sometimes it is useful to run your MCMC chains a little bit longer to improve convergence. Re-starting from the run in previous section, you can use:

```
niter_ad <- 6000
CsurvivalMCMC$run(niter_ad, reset = FALSE)
## |----------------|----------------|----------------|----------------|
## |------------------------------------------------------------------|
```

Then you can extract the matrix of previous MCMC samples augmented with new ones and obtain numerical summaries:

```
more_samples <- as.matrix(CsurvivalMCMC$mvSamples)
samplesSummary(more_samples)
##          Mean Median St.Dev. 95%CI_low 95%CI_upp
## theta 0.3402 0.3382 0.05975    0.2281    0.4632
```

You can check that `more_samples` contains 10000 samples, 4000 from the call to `runMCMC()` plus 6000 additional samples. Note that this only works if you are using the detailed NIMBLE workflow. It does not work with the `nimbleMCMC()` wrapper function.

2.7.5 Reproducibility

If you want your results to be reproducible, you can control the state of R the random number generator with the `setSeed` argument in functions `nimbleMCMC()` and `runMCMC()`. Going back to the animal survival example, you can check that two calls to `nimbleMCMC()` give the same results when `setSeed` is set to the same value. Note that we need to specify a seed for each chain, hence a vector of three components here:

```
# first call to nimbleMCMC()
mcmc.output1 <- nimbleMCMC(code = model,
                           data = my.data,
                           inits = initial.values,
                           niter = 5000,
                           nburnin = 1000,
                           nchains = 3,
                           summary = TRUE,
                           setSeed = c(2024, 2025, 2026))
```

```
## |------------|------------|------------|------------|
## |--------------------------------------------------|
## |------------|------------|------------|------------|
## |--------------------------------------------------|
## |------------|------------|------------|------------|
## |--------------------------------------------------|

# second call to nimbleMCMC()
mcmc.output2 <- nimbleMCMC(code = model,
                          data = my.data,
                          inits = initial.values,
                          niter = 5000,
                          nburnin = 1000,
                          nchains = 3,
                          summary = TRUE,
                          setSeed = c(2024, 2025, 2026))
## |------------|------------|------------|------------|
## |--------------------------------------------------|
## |------------|------------|------------|------------|
## |--------------------------------------------------|
## |------------|------------|------------|------------|
## |--------------------------------------------------|

# outputs from both calls are the same
mcmc.output1$summary$all.chains
##        Mean Median St.Dev. 95%CI_low 95%CI_upp
## theta 0.339 0.3363 0.06141    0.2245    0.4618
mcmc.output2$summary$all.chains
##        Mean Median St.Dev. 95%CI_low 95%CI_upp
## theta 0.339 0.3363 0.06141    0.2245    0.4618
```

Note that to make your workflow reproducible, you need to set the seed not only within the `nimbleMCMC()` call, but also before setting your initial values if you are using a randomized function for that.

2.7.6 Parallelization

To speed up your analyses, you can run MCMC chains in parallel. This is what the package `jagsUI` accomplishes for JAGS users. Here, we use the `parallel` package for parallel computation:

```
library(parallel)
```

First you create a cluster using the total amount of cores you have but one to make sure your computer can go on working:

```
nbcores <- detectCores() - 1
my_cluster <- makeCluster(nbcores)
```

Then you wrap your workflow in a function to be run in parallel:

```
workflow <- function(seed, data) {

  library(nimble)

  model <- nimbleCode({
    # likelihood
    survived ~ dbinom(theta, released)
    # prior
    theta ~ dunif(0, 1)
  })

  set.seed(123) # for reproducibility
  initial.values <- function() list(theta = runif(1,0,1))

  survival <- nimbleModel(code = model,
                          data = data,
                          inits = initial.values())
  Csurvival <- compileNimble(survival)
  survivalMCMC <- buildMCMC(Csurvival)
  CsurvivalMCMC <- compileNimble(survivalMCMC)
```

```r
  samples <- runMCMC(mcmc = CsurvivalMCMC,
                     niter = 5000,
                     nburnin = 1000,
                     setSeed = seed)


  return(samples)
}
```

Now we run the code using `parLapply()`, which uses cluster nodes to execute our workflow:

```r
output <- parLapply(cl = my_cluster,
                    X = c(2022, 666),
                    fun = workflow,
                    data = list(survived = 19, released = 57))
```

In the call to `parLapply`, we specify `X = c(2022, 666)` to ensure reproducibility. We use two values, 2022 and 666 to set the seed in `workflow()`, which means we run two instances of our workflow, or two MCMC chains. Note that we also have a line `set.seed(123)` in the `workflow()` function to ensure reproducibility while drawing randomly initial values.

It's good practice to close the cluster with `stopCluster()` so that processes do not continue to run in the background and slow down other processes:

```r
stopCluster(my_cluster)
```

By inspecting the results, you can see that the object `output` is a list with two components, one for each MCMC chain:

```r
str(output)
## List of 2
##  $ : num [1:4000, 1] 0.393 0.369 0.346 0.346 0.346 ...
```

```
##    ..- attr(*, "dimnames")=List of 2
##    .. ..$ : NULL
##    .. ..$ : chr "theta"
##  $ : num [1:4000, 1] 0.435 0.435 0.435 0.435 0.243 ...
##    ..- attr(*, "dimnames")=List of 2
##    .. ..$ : NULL
##    .. ..$ : chr "theta"
```

Eventually, you can obtain numerical summaries:

```
MCMCsummary(output)
##         mean       sd    2.5%     50%   97.5% Rhat n.eff
## theta 0.3361 0.06148 0.2215 0.3335 0.4594    1  1779
```

2.7.7   Incomplete and incorrect initialization

When you run `nimbleMCMC()` or `nimbleModel()`, you may get warnings
thrown at you by NIMBLE like "This model is not fully initialized" or
"value is NA or NaN even after trying to calculate". This is not necessarily
an error, but it "reflects missing values in model variables" (incomplete
initialization). In this situation, NIMBLE will initialize nodes with NAs
by drawing from priors, and it will work or not. When possible, I try
to initialize all nodes (full initialization). The process can be a bit of a
headache, but it helps understanding the model structure better. Going
back to our animal survival example, let's purposedly forget to provide
an initial value for `theta`:

```
model <- nimbleCode({
  # likelihood
  survived ~ dbinom(theta, released)
  # prior
  theta ~ dunif(0, 1)
})
#initial.values <- list(theta = runif(1,0,1))
```

```
survival <- nimbleModel(code = model,
                        data = list(survived = 19, released = 57))
```

To see which variables are not initialized, we use `initializeInfo()`:

```
# survival$calculate() # gives NA
survival$initializeInfo()
```

Now that we know `theta` was not initialized, we can fix the issue and resume our workflow:

```
survival$theta <- 0.5 # assign initial value to theta
survival$calculate()
## [1] -5.422

Csurvival <- compileNimble(survival)
survivalMCMC <- buildMCMC(Csurvival)
## ===== Monitors ======
## thin = 1: theta
## ===== Samplers ======
## RW sampler (1)
##    - theta
CsurvivalMCMC <- compileNimble(survivalMCMC)

samples <- runMCMC(mcmc = CsurvivalMCMC,
                   niter = 5000,
                   nburnin = 1000)
## |-------------|-------------|-------------|-------------|
## |-------------------------------------------------------|

samplesSummary(samples)
##         Mean Median St.Dev. 95%CI_low 95%CI_upp
## theta 0.3402  0.338 0.06025    0.2277    0.4608
```

When working with larger models, it can happen that NIMBLE's internal simulation produces initial values for different parameters (nodes) that are incompatible with each other and violate certain model assumptions. If we go ahead and run the MCMC on a model where this is the case, a range of different warning messages may appear to indicate the problem. At first, they may not seem very intuitive (e.g. "Log probability is -Inf", "Log probability is NaN", "Slice sampler reached the maximum number of contractions", etc.), but they are signals that you may want to double-check our initialization.

2.7.8   Vectorization

Vectorization is the process of replacing a loop by a vector so that instead of processing a single value at a time, you process a set of values at once. As an example, instead of writing:

```
for(i in 1:n){
  x[i] <- mu + epsilon[i]
}
```

you would write:

```
x[1:n] <- mu + epsilon[1:n]
```

Vectorization can make your code more efficient by manipulating one vector node `x[1:n]` instead of n nodes `x[1]`, ..., `x[n]`. As an example, you may have a look to the vectorized flavor of the binomial distribution written by Pierre Dupont at https://github.com/nimble-dev/nimble SCR/blob/master/nimbleSCR/R/dbinom_vector.R. Note that per now, vectorization only works for deterministic nodes (relationships with <-) but not stochastic ones (relationships with ~).

2.8   Summary

- NIMBLE is an R package that implements for you MCMC algorithms to generate samples from the posterior distribution of model parameters. You only have to specify a likelihood and priors using the BUGS language to apply the Bayes theorem.

- NIMBLE is more than just another MCMC engine. It provides a programming environment so that you have full control when building models and estimating parameters.

- At the core of NIMBLE are `nimbleFunctions` which you can write and compile for faster computation. With `nimbleFunctions` you can mimic basic R syntax, work with vectors and matrices, use logical operators and flow control, and specify many distributions.

- There are two workflows to run NIMBLE. In most situations, `nimbleMCMC()` will serve you well. When you need more control, you can adopt a detailed workflow with `nimbleModel()`, `configureMCMC()`, `buildMCMC()`, `compileNimble()` and `runMCMC()`.

- By having full control of the workflow, you can change default MCMC samplers and even write your own samplers.

2.9   Suggested reading

In this chapter, I have only scratched the surface of what NIMBLE is capable of. Below is a list of pointers that should help you going further with NIMBLE.

- The NIMBLE folks make a lot of useful resources available through the official website https://r-nimble.org.

- The NIMBLE manual https://r-nimble.org/manual/cha-welcome-nimble.html reads like a book with clear explanations and relevant examples.

- You can learn a lot by going through examples at https://r-nimble.org/examples.html and training material from NIMBLE workshops at https://github.com/nimble-training.

- You can keep the NIMBLE cheatsheet https://r-nimble.org/NimbleCheatSheet.pdf near you to remind yourself of the workflow, how to write and use models, or which functions and distributions are available.

- If you have questions, feel free to get in touch with the community of NIMBLE users by emailing the discussion group https://groups .google.com/forum/#!forum/nimble-users. This is a great place to learn, and folks who take the time to answer questions are kind and provide constructive answers. When possible, make sure to provide a reproducible example illustrating your problem.

- You can cite the following reference when using NIMBLE in a publication:

---

de Valpine, P., D. Turek, C. J. Paciorek, C. Anderson-Bergman, D. Temple Lang, and R. Bodik (2017). Programming With Models: Writing Statistical Algorithms for General Model Structures With NIMBLE. *Journal of Computational and Graphical Statistics* **26** (2): 403–13.

- Last, the packages to process NIMBLE results are developed by people whose work should be acknowledged: see Youngflesh [2018] for `MCMCvis`, Turek [2022] for `basicMCMCplots`, Gabry and Mahr [2022] for `bayesplot` and Fernández-i Marín [2016] for `ggmcmc`.

3

Hidden Markov models

3.1 Introduction

In this third chapter, you will learn the basics on Markov models and how to fit them to longitudinal data using NIMBLE. In real life however, individuals may go undetected and their status be unknown. You will also learn how to manipulate the extension of Markov models to hidden states, so-called hidden Markov models.

3.2 Longitudinal data

Let's get back to our survival example, and denote z_i the state of individual i with $z_i = 1$ if alive and $z_i = 0$ if dead. We have a total of $z = \sum_{i=1}^{n} z_i$ survivors out of n released animals with winter survival probability ϕ. Our model so far is a combination of a binomial likelihood and a Beta prior with parameters 1 and 1, which is also a uniform distribution between 0 and 1. It can be written as:

$$z \sim \text{Binomial}(n, \phi) \qquad \text{[likelihood]}$$
$$\phi \sim \text{Beta}(1, 1) \qquad \text{[prior for } \phi\text{]}$$

Because the binomial distribution is just a sum of independent Bernoulli outcomes, you can rewrite this model as:

$$z_i \sim \text{Bernoulli}(\phi), \ i = 1, \dots, N \qquad \text{[likelihood]}$$
$$\phi \sim \text{Beta}(1, 1) \qquad \text{[prior for } \phi\text{]}$$

DOI: 10.1201/9781003244097-3

It is like flipping a coin for each individual and get a survivor with probability ϕ.

In this set up, we consider a single winter. But for many species, we need to collect data on the long term to get a representative estimate of survival. Therefore, what if we had say $T = 5$ winters?

Let us denote $z_{i,t} = 1$ if individual i alive at winter t, and $z_{i,t} = 2$ if dead. Then longitudinal data look like in the table below. Each row is an individual i, and columns are for winters t, or sampling occasions. Variable z is indexed by both i and t, and takes value 1 if individual i is alive in winter t, and 2 otherwise.

```
## # A tibble: 57 x 6
##          id `winter 1` `winter 2` `winter 3` `winter 4`
##       <int>      <int>      <int>      <int>      <int>
##  1     1          1          1          1          1
##  2     2          1          1          1          1
##  3     3          1          1          1          1
##  4     4          1          1          1          1
##  5     5          1          1          1          1
##  6     6          1          1          2          2
##  7     7          1          1          1          1
##  8     8          1          2          2          2
##  9     9          1          1          1          1
## 10    10          1          2          2          2
## # i 47 more rows
## # i 1 more variable: `winter 5` <int>
```

3.3 A Markov model for longitudinal data

Let's think of a model for these data. The objective remains the same, estimating survival. To build this model, we'll make assumptions, go through its components and write down its likelihood. Note that we have already encountered Markov models in Section 1.5.2.

3.3.1   Assumptions

First, we assume that the state of an animal in a given winter, alive or dead, is only dependent on its state the winter before. In other words, the future depends only on the present, not the past. This is a Markov process.

Second, if an animal is alive in a given winter, the probability it survives to the next winter is ϕ. The probability it dies is $1 - \phi$.

Third, if an animal is dead a winter, it remains dead, unless you believe in zombies.

Our Markov process can be represented this way:

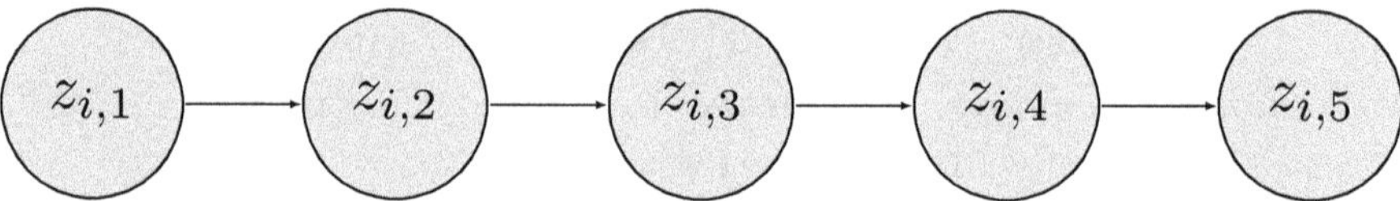

An example of this Markov process is, for example:

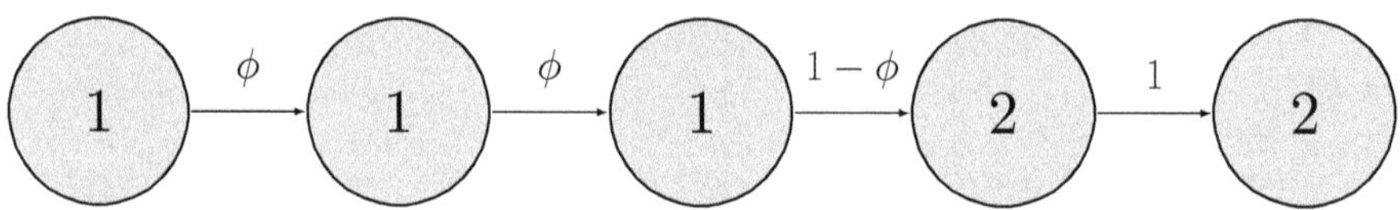

Here the animal remains alive over the first two time intervals ($z_{i,1} = z_{i,2} = z_{i,3} = 1$) with probability ϕ until it dies over the fourth time interval ($z_{i,4} = 2$) with probability $1 - \phi$ then remains dead from then onwards ($z_{i,5} = 2$) with probability 1.

3.3.2   Transition matrix

You might have figured it out already (if not, not a problem), the core of our Markov process is made of transition probabilities between states alive and dead. For example, the probability of transitioning from state alive at $t - 1$ to state alive at t is $\Pr(z_{i,t} = 1|z_{i,t-1} = 1) = \gamma_{1,1}$. It is the survival probability ϕ. The probability of dying over the interval $(t-1, t)$ is $\Pr(z_{i,t} = 2|z_{i,t-1} = 1) = \gamma_{1,2} = 1 - \phi$. Now if an animal is dead at $t - 1$, then $\Pr(z_t = 1|z_{t-1} = 2) = 0$ and $\Pr(z_{i,t} = 2|z_{i,t-1} = 2) = 1$.

We can gather these probabilities of transition between states from one occasion to the next in a matrix, say Γ, which we will call the transition matrix:

$$\Gamma = \left(\begin{array}{cc} \gamma_{1,1} & \gamma_{1,2} \\ \gamma_{2,1} & \gamma_{2,2} \end{array} \right) = \left(\begin{array}{cc} \phi & 1-\phi \\ 0 & 1 \end{array} \right)$$

To try and remember that the states at $t-1$ are in rows, and the states at t are in columns, I will often write:

$$\Gamma = \begin{array}{cc} z_t = 1 & z_t = 2 \\ \left(\begin{array}{cc} \phi & 1-\phi \\ 0 & 1 \end{array} \right) & \begin{array}{l} z_{t-1} = 1 \text{ (alive)} \\ z_{t-1} = 2 \text{ (dead)} \end{array} \end{array}$$

Take the time you need to navigate through this matrix, and get familiar with it. For example, you may start alive at t (first row), then end up dead at $t+1$ (first column) with probability $1 - \phi$.

3.3.3　Initial states

A Markov process has to start somewhere. We need the probabilities of initial states, i.e. the states of an individual at $t = 1$. We will gather the probability of being in each state (alive or 1 and dead or 2) in the first winter in a vector. We will use $\delta = (\Pr(z_{i,1} = 1), \Pr(z_{i,1} = 2))$. For simplicity, we will assume that all individuals are marked and released in the first winter, hence alive when first captured, which means that they are all in state alive or 1 for sure. Therefore, we have $\delta = (1, 0)$.

3.3.4　Likelihood

Now that we have built a Markov model, we need its likelihood to apply the Bayes theorem. The likelihood is the probability of the data, given the model. Here the data are the z, therefore we need $\Pr(z) = \Pr(z_1, z_2, \ldots, z_{T-2}, z_{T-1}, z_T)$.

We're gonna work backward, starting from the last sampling occasion. Using conditional probabilities, the likelihood can be written as the product of the probability of z_T i.e. you're alive or not on the last occasion given your past history, that is the states at previous occasions, times the probability of your past history:

$$\Pr(z) = \Pr(z_T, z_{T-1}, z_{T-2}, \ldots, z_1)$$
$$= \Pr(z_T | z_{T-1}, z_{T-2}, \ldots, z_1) \Pr(z_{T-1}, z_{T-2}, \ldots, z_1)$$

Then because we have a Markov model, we're memory less, that is the probability of next state, here z_T, depends only on the current state, that is z_{T-1}, and not the previous states:

$$\Pr(z) = \Pr(z_T, z_{T-1}, z_{T-2}, \ldots, z_1)$$
$$= \Pr(z_T | z_{T-1}, z_{T-2}, \ldots, z_1) \Pr(z_{T-1}, z_{T-2}, \ldots, z_1)$$
$$= \Pr(z_T | z_{T-1}) \Pr(z_{T-1}, z_{T-2}, \ldots, z_1)$$

You can apply the same reasoning to $T - 1$. First use conditional probabilities:

$$\Pr(z) = \Pr(z_T, z_{T-1}, z_{T-2}, \ldots, z_1)$$
$$= \Pr(z_T | z_{T-1}, z_{T-2}, \ldots, z_1) \Pr(z_{T-1}, z_{T-2}, \ldots, z_1)$$
$$= \Pr(z_T | z_{T-1}) \Pr(z_{T-1}, z_{T-2}, \ldots, z_1)$$
$$= \Pr(z_T | z_{T-1}) \Pr(z_{T-1} | z_{T-2}, \ldots, z_1) \Pr(z_{T-2}, \ldots, z_1)$$

Then apply the Markovian property:

$$\Pr(z) = \Pr(z_T, z_{T-1}, z_{T-2}, \ldots, z_1)$$
$$= \Pr(z_T | z_{T-1}, z_{T-2}, \ldots, z_1) \Pr(z_{T-1}, z_{T-2}, \ldots, z_1)$$
$$= \Pr(z_T | z_{T-1}) \Pr(z_{T-1}, z_{T-2}, \ldots, z_1)$$
$$= \Pr(z_T | z_{T-1}) \Pr(z_{T-1} | z_{T-2}, \ldots, z_1) \Pr(z_{T-2}, \ldots, z_1)$$
$$= \Pr(z_T | z_{T-1}) \Pr(z_{T-1} | z_{T-2}) \Pr(z_{T-2}, \ldots, z_1)$$

And so on up to z_2. You end up with this expression for the likelihood:

$$\begin{aligned}
\Pr(z) &= \Pr(z_T, z_{T-1}, z_{T-2}, \dots, z_1) \\
&= \Pr(z_T | z_{T-1}, z_{T-2}, \dots, z_1) \Pr(z_{T-1}, z_{T-2}, \dots, z_1) \\
&= \Pr(z_T | z_{T-1}) \Pr(z_{T-1}, z_{T-2}, \dots, z_1) \\
&= \Pr(z_T | z_{T-1}) \Pr(z_{T-1} | z_{T-2}, \dots, z_1) \Pr(z_{T-2}, \dots, z_1) \\
&= \Pr(z_T | z_{T-1}) \Pr(z_{T-1} | z_{T-2}) \Pr(z_{T-2}, \dots, z_1) \\
&= \dots \\
&= \Pr(z_T | z_{T-1}) \Pr(z_{T-1} | z_{T-2}) \dots \Pr(z_2 | z_1) \Pr(z_1)
\end{aligned}$$

This is a product of conditional probabilities of states given previous states, and the probability of initial states $\Pr(z_1)$. Using a more compact notation for the product of conditional probabilities, we get:

$$\begin{aligned}
\Pr(z) &= \Pr(z_T, z_{T-1}, z_{T-2}, \dots, z_1) \\
&= \Pr(z_T | z_{T-1}, z_{T-2}, \dots, z_1) \Pr(z_{T-1}, z_{T-2}, \dots, z_1) \\
&= \Pr(z_T | z_{T-1}) \Pr(z_{T-1}, z_{T-2}, \dots, z_1) \\
&= \Pr(z_T | z_{T-1}) \Pr(z_{T-1} | z_{T-2}, \dots, z_1) \Pr(z_{T-2}, \dots, z_1) \\
&= \Pr(z_T | z_{T-1}) \Pr(z_{T-1} | z_{T-2}) \Pr(z_{T-2}, \dots, z_1) \\
&= \dots \\
&= \Pr(z_T | z_{T-1}) \Pr(z_{T-1} | z_{T-2}) \dots \Pr(z_2 | z_1) \Pr(z_1) \\
&= \Pr(z_1) \prod_{t=2}^{T} \Pr(z_t | z_{t-1})
\end{aligned}$$

In the product, you can recognize the transition parameters γ we defined above, so that the likelihood of a Markov model can be written as:

$$\begin{aligned}
\Pr(z) &= \Pr(z_T, z_{T-1}, z_{T-2}, \dots, z_1) \\
&= \Pr(z_1) \prod_{t=2}^{T} \gamma_{z_{t-1}, z_t}
\end{aligned}$$

3.3.5 Example

I realize these calculations are a bit difficult to follow. Let's take an example to fix ideas. Let's assume an animal is alive, alive at time 2 then dies at time 3. We have $z = (1, 1, 2)$. What is the contribution of this animal to the likelihood? Let's apply the formula we just derived:

$$\Pr(z = (1, 1, 2)) = \Pr(z_1 = 1)\; \gamma_{z_1=1,z_2=1}\; \gamma_{z_2=1,z_3=2}$$
$$= 1\, \phi\, (1 - \phi).$$

The probability of having the sequence alive, alive and dead is the probability of being alive first, then to stay alive, eventually to die. The probability of being alive at first occasion being 1, we have that the contribution of this individual to the likelihood is $\phi(1 - \phi)$.

3.4 Bayesian formulation

Before implementing this model in NIMBLE, we provide a Bayesian formulation of our model. We first note that the likelihood is a product of conditional probabilities of binary events (alive or dead). Usually, binary events are associated with the Bernoulli distribution. Here however, we will use its extension to several outcomes (from a coin with two sides to a dice with more than two faces) known as the categorical distribution. The categorical distribution is a multinomial distribution with a single draw. To get a better idea of how the categorical distribution works, let's simulate from it with the `rcat()` function. Consider for example a random value drawn from a categorical distribution with probability 0.1, 0.3 and 0.6. Think of a dice with three faces, face 1 has probability 0.1 of occurring, face 2 probability 0.3 and face 3 has probability 0.6, the sum of these probabilities being 1. We expect to get a 3 more often than a 2 and rarely a 1:

```
rcat(n = 1, prob = c(0.1, 0.3, 0.6))
## [1] 3
```

Alternatively, you can use the `sample()` function and `sample(x = 1:3, size = 1, replace = FALSE, prob = c(0.1, 0.3, 0.6))`. Here is another example in which we sample 20 times in a categorical distribution with probabilities 0.1, 0.1, 0.4, 0.2 and 0.2, hence a dice with 5 faces:

```
rcat(n = 20, prob = c(0.1, 0.1, 0.4, 0.2, 0.2))
##  [1] 5 3 4 5 4 3 4 2 2 3 1 3 5 5 3 4 2 3 2 3
```

In this chapter, you will familiarize yourself with the categorical distribution in binary situations, which should make the transition to more states than just alive and dead smoother in the next chapters.

Initial state is a categorical random variable with probability δ. That is you have a dice with two faces, or a coin, and you have some probability to be alive, and one minus that probability to be dead. Of course, it you want your Markov chain to start, you'd better say it's alive so that δ is just $(1, 0)$:

$$z_1 \sim \text{Categorical}(\delta) \qquad\qquad [\text{likelihood}, t = 1]$$

Now the main part is the dynamic of the states. The state z_t at t depends only on the known state z_{t-1} at $t-1$ and is a categorical random variable which probabilities are given by row z_{t-1} of the transition matrix $\Gamma = \gamma_{z_{t-1},z_t}$:

$$z_1 \sim \text{Categorical}(\delta) \qquad\qquad [\text{likelihood}, t = 1]$$
$$z_t|z_{t-1} \sim \text{Categorical}(\gamma_{z_{t-1},z_t}) \qquad\qquad [\text{likelihood}, t > 1]$$

For example, if individual i is alive over $(t - 1, t)$ i.e. $z_{t-1} = 1$, we need the first row in Γ,

$$\Gamma = \begin{pmatrix} \phi & 1 - \phi \\ 0 & 1 \end{pmatrix}$$

that is $\gamma_{z_{t-1}=1,z_t} = (\phi, 1 - \phi)$ and $z_t|z_{t-1} = 1 \sim \text{Categorical}((\phi, 1 - \phi))$.

Otherwise, if individual i dies over $(t-1, t)$ i.e. $z_{t-1} = 2$, we need the second row in Γ:

$$\Gamma = \begin{pmatrix} \phi & 1-\phi \\ 0 & 1 \end{pmatrix}$$

that is $\gamma_{z_{t-1}=2,z_t} = (0,1)$ and $z_t | z_{t-1} = 2 \sim \text{Categorical}((0,1))$ (if the individual is dead, it remains dead with probability 1).

We also need a prior on survival. Without surprise, we will use a uniform distribution between 0 and 1, which is also a Beta distribution with parameters 1 and 1. Overall our model is:

$$\begin{aligned} z_1 &\sim \text{Categorical}(\delta) && \text{[likelihood, } t = 1] \\ z_t | z_{t-1} &\sim \text{Categorical}(\gamma_{z_{t-1},z_t}) && \text{[likelihood, } t > 1] \\ \phi &\sim \text{Beta}(1,1) && \text{[prior for } \phi] \end{aligned}$$

3.5   NIMBLE implementation

How to implement in NIMBLE the Markov model we just built? We need to put in place a few bricks before running our model. Let's start with the prior on survival, the vector of initial state probabilities and the transition matrix:

```r
markov.survival <- nimbleCode({
  phi ~ dunif(0, 1) # prior
  delta[1] <- 1            # Pr(alive t = 1) = 1
  delta[2] <- 0            # Pr(dead t = 1) = 0
  gamma[1,1] <- phi        # Pr(alive t -> alive t+1)
  gamma[1,2] <- 1 - phi    # Pr(alive t -> dead t+1)
  gamma[2,1] <- 0          # Pr(dead t -> alive t+1)
  gamma[2,2] <- 1          # Pr(dead t -> dead t+1)
  ...
```

Alternatively, you can define vectors and matrices in NIMBLE like you would do it in R. You can write:

```r
markov.survival <- nimbleCode({
  # prior
  phi ~ dunif(0, 1)
  # vector of initial state probabilities
  delta[1:2] <- c(1, 0)
  # transition matrix
  gamma[1:2,1:2] <- matrix( c(phi, 0, 1 - phi, 1), nrow = 2)
...
```

Now there are two important dimensions to our model, along which we need to repeat tasks, namely individual and time. As for time, we describe the successive events of survival using the categorical distribution dcat(), say for individual i:

```r
z[i,1] ~ dcat(delta[1:2])            # t = 1
z[i,2] ~ dcat(gamma[z[i,1], 1:2])    # t = 2
z[i,3] ~ dcat(gamma[z[i,2], 1:2])    # t = 3
...
z[i,T] ~ dcat(gamma[z[i,T-1], 1:2])  # t = T
```

There is a more efficient way to write this piece of code by using a for loop, that is a sequence of instructions that we repeat. Here, we condense the previous code into:

```r
z[i,1] ~ dcat(delta[1:2])                   # t = 1
for (t in 2:T){ # loop over time t
  z[i,t] ~ dcat(gamma[z[i,t-1], 1:2]) # t = 2,...,T
}
```

Now we just need to do the same for all individuals. We use another loop:

```r
for (i in 1:N){ # loop over individual i
  z[i,1] ~ dcat(delta[1:2]) # t = 1
  for (j in 2:T){ # loop over time t
    z[i,j] ~ dcat(gamma[z[i,j-1], 1:2]) # t = 2,...,T
  } # t
} # i
```

Putting everything together, the NIMBLE code for our Markov model is:

```r
markov.survival <- nimbleCode({
  phi ~ dunif(0, 1) # prior
  delta[1] <- 1            # Pr(alive t = 1) = 1
  delta[2] <- 0            # Pr(dead t = 1) = 0
  gamma[1,1] <- phi        # Pr(alive t -> alive t+1)
  gamma[1,2] <- 1 - phi    # Pr(alive t -> dead t+1)
  gamma[2,1] <- 0          # Pr(dead t -> alive t+1)
  gamma[2,2] <- 1          # Pr(dead t -> dead t+1)
  # likelihood
  for (i in 1:N){ # loop over individual i
    z[i,1] ~ dcat(delta[1:2]) # t = 1
    for (j in 2:T){ # loop over time t
      z[i,j] ~ dcat(gamma[z[i,j-1], 1:2]) # t = 2,...,T
    } # t
  } # i
})
```

Note that in this example, δ is used as a placeholder for more complex models we will build in chapters to come. Here, you could simply write `z[i,1] <- 1`.

Note also that we could replace `dcat()` by `dbern()` everywhere in the code because we have binary events alive/dead. Would it make any difference? Although `dcat()` uses less efficient samplers than `dbern()`, `dcat()` is convenient for model building to accommodate more than two outcomes, a feature that will become handy in the next chapters.

Now we're ready to resume our NIMBLE workflow. First we read in data. The code I used to simulate the z with survival $\phi = 0.8$ is as follows:

```r
# 1 = alive, 2 = dead
nind <- 57
nocc <- 5
phi <- 0.8 # survival probability
delta <- c(1,0) # (Pr(alive at t = 1), Pr(dead at t = 1))
Gamma <- matrix(NA, 2, 2) # transition matrix
Gamma[1,1] <- phi       # Pr(alive t -> alive t+1)
Gamma[1,2] <- 1 - phi   # Pr(alive t -> dead t+1)
Gamma[2,1] <- 0         # Pr(dead t -> alive t+1)
Gamma[2,2] <- 1         # Pr(dead t -> dead t+1)
z <- matrix(NA, nrow = nind, ncol = nocc)
set.seed(2022)
for (i in 1:nind){
  z[i,1] <- rcat(n = 1, prob = delta) # 1 for sure
  for (t in 2:nocc){
    z[i,t] <- rcat(n = 1, prob = Gamma[z[i,t-1],1:2])
  }
}
head(z)
##      [,1] [,2] [,3] [,4] [,5]
## [1,]    1    1    1    1    1
## [2,]    1    1    1    1    1
## [3,]    1    1    1    1    1
## [4,]    1    1    1    1    2
## [5,]    1    1    1    1    1
## [6,]    1    1    2    2    2
```

Note the resemblance with the model NIMBLE code above. Because we have loops and indices that do not change, we use constants as explained in Section 2.3:

```r
my.data <- list(z = z)
my.constants <- list(N = 57, T = 5)
```

We also specify initial values for survival with a function:

```
initial.values <- function() list(phi = runif(1,0,1))
initial.values()
## $phi
## [1] 0.6503
```

There is a single parameter to monitor:

```
parameters.to.save <- c("phi")
parameters.to.save
## [1] "phi"
```

We run 2 chains with 5000 iterations including 1000 iterations as burnin:

```
n.iter <- 5000
n.burnin <- 1000
n.chains <- 2
```

Let's run NIMBLE:

```
mcmc.output <- nimbleMCMC(code = markov.survival,
                          constants = my.constants,
                          data = my.data,
                          inits = initial.values,
                          monitors = parameters.to.save,
                          niter = n.iter,
                          nburnin = n.burnin,
                          nchains = n.chains)
```

Let's calculate the usual posterior numerical summaries for survival:

```
MCMCsummary(mcmc.output, round = 2)
##      mean   sd 2.5%  50% 97.5% Rhat n.eff
## phi 0.79 0.03 0.73 0.79  0.85    1  1751
```

Posterior mean and median are close to 0.8. This is fortunate since the data was simulated with (actual) survival $\phi = 0.8$.

3.6　Hidden Markov models

3.6.1　Capture-recapture data

> Unfortunately, the data with alive and dead states is the data we wish we had. In real life, animals cannot be monitored exhaustively, like humans in a medical trial. This is why we use capture-recapture protocols, in which animals are captured, individually marked, and released alive. Then, these animals may be detected again, or go undetected. Whenever animals go undetected, it might be that they were alive but missed, or because they were dead and therefore could not be detected. This issue is usually referred to as that of imperfect detection. As a consequence of imperfect detection, the Markov process for survival is only partially observed: You know an animal is alive when you detect it, but when an animal goes undetected, whether it is alive or dead is unknown to you. This is where hidden Markov models (HMMs) come in.

Let's get back to the data we had in the previous section. The truth is in z which contains the fate of all individuals with $z = 1$ for alive, and $z = 2$ for dead:

```
## # A tibble: 57 x 6
##         id `winter 1` `winter 2` `winter 3` `winter 4`
##      <int>      <int>      <int>      <int>      <int>
## 1     1          1          1          1          1
## 2     2          1          1          1          1
## 3     3          1          1          1          1
## 4     4          1          1          1          1
## 5     5          1          1          1          1
## 6     6          1          1          2          2
```

```
## 7     7          1          1          1          1
## 8     8          1          2          2          2
## 9     9          1          1          1          1
## 10    10         1          2          2          2
## # i 47 more rows
## # i 1 more variable: `winter 5` <int>
```

Unfortunately, we have only partial access to z. What we do observe is y the detections and non-detections. How are z and y connected?

The easiest connection is with dead animals which go undetected for sure. Therefore, when an animal is dead i.e. $z = 2$, it cannot be detected, therefore $y = 0$:

```
## # A tibble: 57 x 6
##          id `winter 1` `winter 2` `winter 3` `winter 4`
##       <int>      <int>      <int>      <int>      <int>
## 1     1          1          1          1          1
## 2     2          1          1          1          1
## 3     3          1          1          1          1
## 4     4          1          1          1          1
## 5     5          1          1          1          1
## 6     6          1          1          0          0
## 7     7          1          1          1          1
## 8     8          1          0          0          0
## 9     9          1          1          1          1
## 10    10         1          0          0          0
## # i 47 more rows
## # i 1 more variable: `winter 5` <int>
```

Now alive animals may be detected or not. If an animal is alive $z = 1$, it is detected $y = 1$ with probability p or not $y = 0$ with probability $1 - p$. In our example, first detection coincides with first winter for all individuals.

```
## # A tibble: 57 x 6
##          id `winter 1` `winter 2` `winter 3` `winter 4`
##       <int>      <dbl>      <dbl>      <dbl>      <dbl>
## 1     1          1          0          0          0
## 2     2          1          0          1          0
```

```
## 3     3             1          0          0          0
## 4     4             1          1          1          1
## 5     5             1          1          1          1
## 6     6             1          0          0          0
## 7     7             1          0          1          1
## 8     8             1          1          1          1
## 9     9             1          1          1          1
## 10    10            1          1          0          0
## # i 47 more rows
## # i 1 more variable: `winter 5` <dbl>
```

Compare with the previous z table. Some 1's for alive have become 0's for
non-detection, other 1's for alive have remained 1's for detection. This
y table is what we observe in real life. I hope I have convinced you that
to make the connection between observations, the y, and true states,
the z, we need to describe how observations are made (or emitted in
the HMM terminology) from the states.

3.6.2   Observation matrix

The novelty in HMMs is the link between observations and states. This
link is made through observation probabilities. For example, the proba-
bility of detecting an animal i at t given it is alive at t is $\Pr(y_{i,t} = 2 | z_{i,t} = 1) = \omega_{1,2}$. It is the detection probability p. If individual i is dead at t,
then it is missed for sure, and $\Pr(y_{i,t} = 1 | z_{i,t} = 2) = \omega_{2,1} = 1$.

We can gather these observation probabilities into an observation matrix Ω. In rows we have the states alive $z = 1$ and dead $z = 2$, while in
columns we have the observations non-detected $y = 1$ and detected
$y = 2$ (previously coded 0 and 1 respectively):

$$\Omega = \begin{pmatrix} \omega_{1,1} & \omega_{1,2} \\ \omega_{2,1} & \omega_{2,2} \end{pmatrix} = \begin{pmatrix} 1-p & p \\ 1 & 0 \end{pmatrix}$$

In survival models we will use throughout this book, we condition
the fate of individuals on first detection, which boils down to set the
corresponding detection probability to 1.

The observation matrix is:

$$\Omega = \begin{array}{c} \quad y_t = 1 \text{ (non-detected)} \quad y_t = 2 \text{ (detected)} \\ \begin{pmatrix} 1-p & p \\ 1 & 0 \end{pmatrix} \end{array} \begin{array}{l} z_t = 1 \text{ (alive)} \\ z_t = 2 \text{ (dead)} \end{array}$$

3.6.3　Hidden Markov model

Our hidden Markov model can be represented this way:

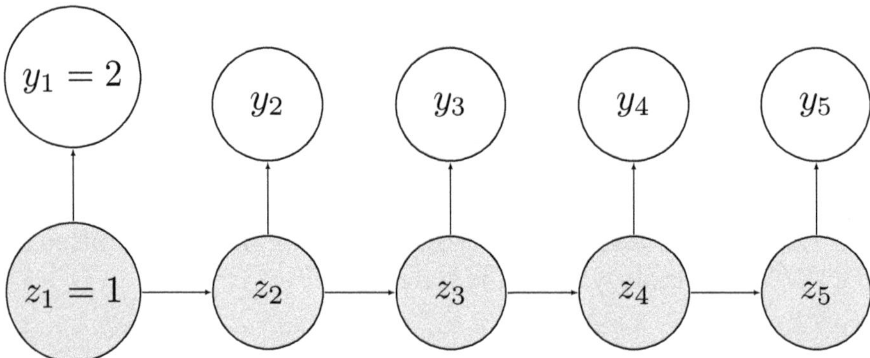

States z are in gray. Observations y are in white. All individuals are first captured in the first winter $t = 1$ and are therefore all alive $z_1 = 1$ and detected $y_1 = 2$.

> A hidden Markov model is just two time series running in parallel. One for the states with the Markovian property, and the other for the observations generated from the states. HMM are a special case of state-space models in which latent states are discrete.

Have a look to the example below, in which an individual is detected at first sampling occasion, detected again, then missed for the rest of the study. While on occasion $t = 3$ that individual was alive $z_3 = 1$ and went undetected $y_3 = 1$, on occasions $t = 4$ and $t = 5$ it went undetected $y_4 = y_5 = 1$ because it was dead $z_4 = z_5 = 2$. Because we condition on first detection, the link between state and observation at $t = 1$ is deterministic and $p = 1$.

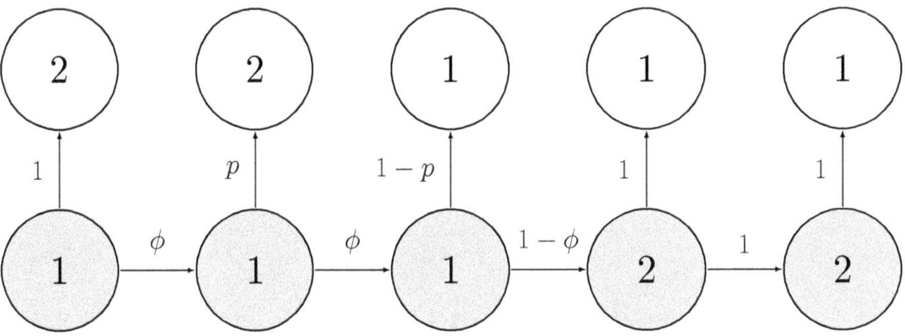

3.6.4 Likelihood

In the Bayesian framework, we usually work with the so-called complete likelihood, that is the probability of the observed data y and the latent states z given the parameters of our model, here the survival and detection probabilities ϕ and p. The complete likelihood for individual i is:

$$\Pr(y_i, z_i) = \Pr(y_{i,1}, y_{i,2}, \ldots, y_{i,T}, z_{i,1}, z_{i,2}, \ldots, z_{i,T})$$

Using the definition of a conditional probability, we have:

$$\begin{aligned}
\Pr(y_i, z_i) &= \Pr(y_{i,1}, y_{i,2}, \ldots, y_{i,T}, z_{i,1}, z_{i,2}, \ldots, z_{i,T}) \\
&= \Pr(y_{i,1}, y_{i,2}, \ldots, y_{i,T} | z_{i,1}, z_{i,2}, \ldots, z_{i,T}) \\
&\quad \times \Pr(z_{i,1}, z_{i,2}, \ldots, z_{i,T})
\end{aligned}$$

Then by using the independence of the y conditional on the z, and the likelihood of a Markov chain, we get that:

$$\begin{aligned}
\Pr(y_i, z_i) &= \Pr(y_{i,1}, y_{i,2}, \ldots, y_{i,T}, z_{i,1}, z_{i,2}, \ldots, z_{i,T}) \\
&= \Pr(y_{i,1}, y_{i,2}, \ldots, y_{i,T} | z_{i,1}, z_{i,2}, \ldots, z_{i,T}) \\
&\quad \times \Pr(z_{i,1}, z_{i,2}, \ldots, z_{i,T}) \\
&= \left(\prod_{t=1}^{T} \Pr(y_{i,t} | z_{i,t}) \right) \left(\Pr(z_{i,1}) \prod_{t=2}^{T} \Pr(z_{i,t} | z_{i,t-1}) \right)
\end{aligned}$$

Finally, by recognizing the observation and transition probabilities, we have that the complete likelihood for individual i is:

$$\begin{aligned}
\Pr(y_i, z_i) &= \Pr(y_{i,1}, y_{i,2}, \ldots, y_{i,T}, z_{i,1}, z_{i,2}, \ldots, z_{i,T}) \\
&= \Pr(y_{i,1}, y_{i,2}, \ldots, y_{i,T} | z_{i,1}, z_{i,2}, \ldots, z_{i,T}) \\
&\quad \times \Pr(z_{i,1}, z_{i,2}, \ldots, z_{i,T}) \\
&= \left(\prod_{t=1}^{T} \omega_{z_{i,t}, y_{i,t}} \right) \left(\Pr(z_{i,1}) \prod_{t=2}^{T} \gamma_{z_{i,t-1}, z_{i,t}} \right)
\end{aligned}$$

To obtain the complete likelihood of the whole dataset, we need to multiply this individual likelihood for each animal $\prod_{i=1}^{N} \Pr(y_i, z_i)$. When several individuals have the same contribution, calculating their individual contribution only once can greatly reduce the computational burden, as illustrated in Section 3.9.

The Bayesian approach with MCMC methods allows treating the latent states $z_{i,t}$ as if they were parameters, and to be estimated as such. However, the likelihood is rather complex with a large number of latent states $z_{i,t}$, which comes with computational costs and slow mixing. There are situations where the latent states are the focus of ecological inference and need to be estimated (see Suggested reading below). However, if not needed, you might want to get rid of the latent states and rely on the so-called marginal likelihood. By doing so, you can avoid sampling the latent states, focus on the ecological parameters, and often speeds up computations and improves mixing as shown in Section 3.8. Actually, you can even estimate the latent states afterwards, as illustrated in Section 3.10.

3.7 Fitting HMM with NIMBLE

If we denote *first* the time of first detection, then our model so far is written as follows:

$$z_{\text{first}} \sim \text{Categorical}(1, \delta) \qquad [\text{likelihood}]$$
$$z_t | z_{t-1} \sim \text{Categorical}(1, \gamma_{z_{t-1}, z_t}) \qquad [\text{likelihood, t>first}]$$
$$y_t | z_t \sim \text{Categorical}(1, \omega_{z_t}) \qquad [\text{likelihood, t>first}]$$
$$\phi \sim \text{Beta}(1, 1) \qquad [\text{prior for } \phi]$$
$$p \sim \text{Beta}(1, 1) \qquad [\text{prior for } p]$$

It has an observation layer for the y's, conditional on the z's. We also consider uniform priors for the detection and survival probabilities. How to implement this model in NIMBLE?

We start with priors for survival and detection probabilities:

```r
hmm.survival <- nimbleCode({
  phi ~ dunif(0, 1) # prior survival
  p ~ dunif(0, 1) # prior detection
...
```

Then we define initial states, transition and observation matrices:

```r
...
  # parameters
  delta[1] <- 1            # Pr(alive t = first) = 1
  delta[2] <- 0            # Pr(dead t = first) = 0
  gamma[1,1] <- phi        # Pr(alive t -> alive t+1)
  gamma[1,2] <- 1 - phi    # Pr(alive t -> dead t+1)
  gamma[2,1] <- 0          # Pr(dead t -> alive t+1)
  gamma[2,2] <- 1          # Pr(dead t -> dead t+1)
  omega[1,1] <- 1 - p      # Pr(alive t -> non-detected t)
  omega[1,2] <- p          # Pr(alive t -> detected t)
  omega[2,1] <- 1          # Pr(dead t -> non-detected t)
  omega[2,2] <- 0          # Pr(dead t -> detected t)
...
```

Then the likelihood:

```
...
    # likelihood
    for (i in 1:N){
    z[i,1] ~ dcat(delta[1:2])
    for (j in 2:T){
      z[i,j] ~ dcat(gamma[z[i,j-1], 1:2])
      y[i,j] ~ dcat(omega[z[i,j], 1:2])
    }
  }
})
```

The loop over time for each individual `for (j in 2:T){}` starts after the first time individuals are detected (this is time 2 for all of them here), because we work conditional on the first detection.

Overall, the code looks like:

```
hmm.survival <- nimbleCode({
  phi ~ dunif(0, 1) # prior survival
  p ~ dunif(0, 1) # prior detection
  # likelihood
  delta[1] <- 1          # Pr(alive t = first) = 1
  delta[2] <- 0          # Pr(dead t = first) = 0
  gamma[1,1] <- phi      # Pr(alive t -> alive t+1)
  gamma[1,2] <- 1 - phi  # Pr(alive t -> dead t+1)
  gamma[2,1] <- 0        # Pr(dead t -> alive t+1)
  gamma[2,2] <- 1        # Pr(dead t -> dead t+1)
  omega[1,1] <- 1 - p    # Pr(alive t -> non-detected t)
  omega[1,2] <- p        # Pr(alive t -> detected t)
  omega[2,1] <- 1        # Pr(dead t -> non-detected t)
  omega[2,2] <- 0        # Pr(dead t -> detected t)
  for (i in 1:N){
    z[i,1] ~ dcat(delta[1:2])
    for (j in 2:T){
      z[i,j] ~ dcat(gamma[z[i,j-1], 1:2])
```

```r
      y[i,j] ~ dcat(omega[z[i,j], 1:2])
    }
  }
})
```

Now we specify the constants:

```r
my.constants <- list(N = nrow(y), T = 5)
my.constants
## $N
## [1] 57
##
## $T
## [1] 5
```

The data are made of 0's for non-detections and 1's for detections. To simulate the y, here is the code I used:

```r
set.seed(2022) # for reproducibility
nocc <- 5 # nb of winters or sampling occasions
nind <- 57 # nb of animals
p <- 0.6 # detection prob
phi <- 0.8 # survival prob
# Vector of initial states probabilities
delta <- c(1,0) # all individuals are alive in first winter
# Transition matrix
Gamma <- matrix(NA, 2, 2)
Gamma[1,1] <- phi        # Pr(alive t -> alive t+1)
Gamma[1,2] <- 1 - phi    # Pr(alive t -> dead t+1)
Gamma[2,1] <- 0          # Pr(dead t -> alive t+1)
Gamma[2,2] <- 1          # Pr(dead t -> dead t+1)
# Observation matrix
Omega <- matrix(NA, 2, 2)
Omega[1,1] <- 1 - p      # Pr(alive t -> non-detected t)
Omega[1,2] <- p          # Pr(alive t -> detected t)
```

```r
Omega[2,1] <- 1          # Pr(dead t -> non-detected t)
Omega[2,2] <- 0          # Pr(dead t -> detected t)
# Matrix of states
z <- matrix(NA, nrow = nind, ncol = nocc)
y <- z
y[,1] <- 2 # all individuals are detected in first winter,
           # as we condition on first detection
for (i in 1:nind){
  z[i,1] <- rcat(n = 1, prob = delta) # 1 for sure
  for (t in 2:nocc){
    # state at t given state at t-1
    z[i,t] <- rcat(n = 1, prob = Gamma[z[i,t-1],1:2])
    # observation at t given state at t
    y[i,t] <- rcat(n = 1, prob = Omega[z[i,t],1:2])
  }
}
y
y <- y - 1 # non-detection = 0, detection = 1
```

To use the categorical distribution, we need to code 1's and 2's. We simply add 1 to get the correct format, that is $y = 1$ for non-detection and $y = 2$ for detection:

```r
my.data <- list(y = y + 1)
```

Now let's write a function for the initial values:

```r
zinits <- y + 1 # non-detection -> alive
zinits[zinits == 2] <- 1 # dead -> alive
initial.values <- function() list(phi = runif(1,0,1),
                                   p = runif(1,0,1),
                                   z = zinits)
```

As initial values for the latent states, we assumed that whenever an individual was non-detected, it was alive, with `zinits <- y + 1`, and we make sure dead individuals are alive with `zinits[zinits == 2] <- 1`.

We specify the parameters we'd like to monitor:

```r
parameters.to.save <- c("phi", "p")
parameters.to.save
## [1] "phi" "p"
```

We provide MCMC details:

```r
n.iter <- 5000
n.burnin <- 1000
n.chains <- 2
```

At last, we're ready to run NIMBLE:

```r
start_time <- Sys.time()
mcmc.output <- nimbleMCMC(code = hmm.survival,
                          constants = my.constants,
                          data = my.data,
                          inits = initial.values,
                          monitors = parameters.to.save,
                          niter = n.iter,
                          nburnin = n.burnin,
                          nchains = n.chains)
end_time <- Sys.time()
end_time - start_time
```

```
## Time difference of 37.31 secs
```

We can have a look to numerical summaries:

```r
MCMCsummary(mcmc.output, round = 2)
##      mean   sd 2.5%  50% 97.5% Rhat n.eff
## p    0.61 0.06 0.50 0.61  0.72    1   740
## phi  0.75 0.04 0.67 0.75  0.83    1   805
```

The estimates for survival and detection are close to true survival $\phi = 0.8$ and detection $p = 0.6$ with which we simulated the data.

3.8 Marginalization

In some situations, you will not be interested in inferring the hidden states $z_{i,t}$, so why bother estimating them? The good news is that you can get rid of the states, so that the marginal likelihood is a function of survival and detection probabilities ϕ and p only.

3.8.1 Brute-force approach

Using the formula of total probability, you get the marginal likelihood by summing over all possible states in the complete likelihood:

$$
\begin{aligned}
\Pr(y_i) &= \Pr(y_{i,1}, y_{i,2}, \ldots, y_{i,T}) \\
&= \sum_{z_i} \Pr(y_i, z_i) \\
&= \sum_{z_{i,1}} \cdots \sum_{z_{i,T}} \Pr(y_{i,1}, y_{i,2}, \ldots, y_{i,T}, z_{i,1}, z_{i,2}, \ldots, z_{i,T})
\end{aligned}
$$

Going through the same steps as for deriving the complete likelihood, we obtain the marginal likelihood:

$$
\Pr(y_i) = \sum_{z_{i,1}} \cdots \sum_{z_{i,T}} \left(\prod_{t=1}^{T} \omega_{z_{i,t}, y_{i,t}} \right) \left(\Pr(z_{i,1}) \prod_{t=2}^{T} \gamma_{z_{i,t-1}, z_{i,t}} \right)
$$

Let's go through an example. Let's imagine we have $T = 3$ winters, and we'd like to write the likelihood for an individual having the encounter history detected, detected then non-detected. Remember that non-detected is coded 1 and detected is coded 2, while alive is coded 1 and dead is coded 2. We need to calculate $\Pr(y_1 = 2, y_2 = 2, y_3 = 1)$ which, according to the formula above, is given by:

$$
\begin{aligned}
\Pr(y_1 = 2, y_2 = 2, y_3 = 1) = \sum_{i=1}^{2} \sum_{j=1}^{2} \sum_{k=1}^{2} & \Pr(y_1 = 2 | z_1 = i) \\
& \Pr(y_2 = 2 | z_2 = j) \Pr(y_3 = 1 | z_3 = k) \\
& \Pr(z_1 = i) \Pr(z_2 = j | z_1 = i) \\
& \Pr(z_3 = k | z_2 = j)
\end{aligned}
$$

Expliciting all the sums in $\Pr(y_1 = 2, y_2 = 2, y_3 = 1)$, we get the long and ugly expression:

$$\begin{aligned}
\Pr(y_1 = 2, y_2 = 2, y_3 = 1) =& \\
\Pr(y_1 = 2|z_1 = 1)\,&\Pr(y_2 = 2|z_2 = 1)\,\Pr(y_3 = 1|z_3 = 1)\times \\
&\Pr(z_1 = 1)\,\Pr(z_2 = 1|z_1 = 1)\,\Pr(z_3 = 1|z_2 = 1)+ \\
\Pr(y_1 = 2|z_1 = 2)\,&\Pr(y_2 = 2|z_2 = 1)\,\Pr(y_3 = 1|z_3 = 1)\times \\
&\Pr(z_1 = 2)\,\Pr(z_2 = 1|z_1 = 2)\,\Pr(z_3 = 1|z_2 = 1)+ \\
\Pr(y_1 = 2|z_1 = 1)\,&\Pr(y_2 = 2|z_2 = 2)\,\Pr(y_3 = 1|z_3 = 1)\times \\
&\Pr(z_1 = 1)\,\Pr(z_2 = 2|z_1 = 1)\,\Pr(z_3 = 1|z_2 = 2)+ \\
\Pr(y_1 = 2|z_1 = 2)\,&\Pr(y_2 = 2|z_2 = 2)\,\Pr(y_3 = 1|z_3 = 1)\times \\
&\Pr(z_1 = 2)\,\Pr(z_2 = 2|z_1 = 2)\,\Pr(z_3 = 1|z_2 = 2)+ \\
\Pr(y_1 = 2|z_1 = 1)\,&\Pr(y_2 = 2|z_2 = 1)\,\Pr(y_3 = 1|z_3 = 2)\times \\
&\Pr(z_1 = 1)\,\Pr(z_2 = 1|z_1 = 1)\,\Pr(z_3 = 2|z_2 = 1)+ \\
\Pr(y_1 = 2|z_1 = 2)\,&\Pr(y_2 = 2|z_2 = 1)\,\Pr(y_3 = 1|z_3 = 2)\times \\
&\Pr(z_1 = 2)\,\Pr(z_2 = 1|z_1 = 2)\,\Pr(z_3 = 2|z_2 = 1)+ \\
\Pr(y_1 = 2|z_1 = 1)\,&\Pr(y_2 = 2|z_2 = 2)\,\Pr(y_3 = 1|z_3 = 2)\times \\
&\Pr(z_1 = 1)\,\Pr(z_2 = 2|z_1 = 1)\,\Pr(z_3 = 2|z_2 = 2)+ \\
\Pr(y_1 = 2|z_1 = 2)\,&\Pr(y_2 = 2|z_2 = 2)\,\Pr(y_3 = 1|z_3 = 2)\times \\
&\Pr(z_1 = 2)\,\Pr(z_2 = 2|z_1 = 2)\,\Pr(z_3 = 2|z_2 = 2)
\end{aligned}$$

You can simplify this expression by noticing that i) all individuals are alive for sure when marked and released in first winter, or $\Pr(z_1 = 2) = 0$ and ii) dead individuals are non-detected for sure, or $\Pr(y_t = 2|z_t = 2) = 0$, which lead to:

$$\begin{aligned}
\Pr(y_1 = 2, y_2 = 2, y_3 = 1) =& \\
\Pr(y_1 = 2|z_1 = 1)\,&\Pr(y_2 = 2|z_2 = 1)\,\Pr(y_3 = 1|z_3 = 1)\times \\
&\Pr(z_1 = 1)\,\Pr(z_2 = 1|z_1 = 1)\,\Pr(z_3 = 1|z_2 = 1)+ \\
\Pr(y_1 = 2|z_1 = 1)\,&\Pr(y_2 = 2|z_2 = 1)\,\Pr(y_3 = 1|z_3 = 2)\times \\
&\Pr(z_1 = 1)\,\Pr(z_2 = 1|z_1 = 1)\,\Pr(z_3 = 2|z_2 = 1)
\end{aligned}$$

Because all individuals are captured in first winter, or $\Pr(y_1 = 2 | z_1 = 1) = 1$, we get:

$$\Pr(y_1 = 2, y_2 = 2, y_3 = 1) = 1(1-p) \times 1\phi\phi + 1p1 \times 1\phi(1-\phi)$$

You end up with $\Pr(y_1 = 2, y_2 = 2, y_3 = 1) = \phi p(1 - p\phi)$.

The latent states are no longer involved in the likelihood for this individual. However, even on a rather simple example, the marginal likelihood is quite complex to evaluate because it involves many operations. If T is the length of our encounter histories and N is the number of hidden states (two for alive and dead, but we will deal with more states in some chapters to come), then we need to calculate the sum of N^T terms (the sums in the formula above), each of which has two products of T factors (the products in the formula above), hence $2TN^T$ calculations in total. You can check that in the simple example above, we have $T^N = 2^3 = 8$ terms that are summed, each of which is a product of $2T = 2 \times 3 = 6$ terms. This means that the number of operations increases exponentially as the number of states increases. In most cases, this complexity precludes using this method to get rid of the states. Fortunately, we have another algorithm in the HMM toolbox that is useful to calculate the marginal likelihood efficiently.

3.8.2   Forward algorithm

In the brute-force approach, some products are computed several times to calculate the marginal likelihood. What if we could store these products and use them later while computing the probability of the observation sequence? This is precisely what the forward algorithm does.

We introduce $\alpha_t(j)$, the probability for the latent state z of being in state j at t after seeing the first j observations $y_1, \ldots, y_t$, that is $\alpha_t(j) = \Pr(y_1, \ldots, y_t, z_t = j)$.

Using the law of total probability, we can write the marginal likelihood as a function of $\alpha_T(j)$, namely we have $\Pr(y) =$
$$\sum_{j=1}^{N} \Pr(y_1, \ldots, y_t, z_t = j) = \sum_{j=1}^{N} \alpha_T(j).$$

How to calculate the $\alpha_T(j)$s? This is where the magic of the forward algorithm happens. We use a recurrence relationship that saves us many computations.

The recurrence states that:

$$\alpha_t(j) = \sum_{i=1}^{N} \alpha_{t-1}(i)\gamma_{i,j}\omega_{j,y_t}$$

How to obtain this recurrence? First, using the law of total probability with z_{t-1}, we have that:

$$\alpha_t(j) = \sum_{i=1}^{N} \Pr(y_1, \ldots, y_t, z_{t-1} = i, z_t = j)$$

Second, using conditional probabilities, we get:

$$\alpha_t(j) = \sum_{i=1}^{N} \Pr(y_t | z_{t-1} = i, z_t = j, y_1, \ldots, y_t)$$
$$\times \Pr(z_{t-1} = i, z_t = j, y_1, \ldots, y_t)$$

Third, using conditional probabilities again, on the second term of the product, we get:

$$\alpha_t(j) = \sum_{i=1}^{N} \Pr(y_t | z_{t-1} = i, z_t = j, y_1, \ldots, y_t) \times$$
$$\Pr(z_t = j | z_{t-1} = i, y_1, \ldots, y_t) \Pr(z_{t-1} = i, y_1, \ldots, y_t)$$

which, using conditional independence, simplifies into:

$$\alpha_t(j) = \sum_{i=1}^{N} \Pr(y_t | z_t = j) \Pr(z_t = j | z_{t-1} = i) \Pr(z_{t-1} = i, y_1, \ldots, y_t)$$

Recognizing that $\Pr(y_t | z_t = j) = \omega_{j,y_t}$, $\Pr(z_t = j | z_{t-1} = i) = \gamma_{i,j}$ and $\Pr(z_{t-1} = i, y_1, \ldots, y_t) = \alpha_{t-1}(i)$, we obtain the recurrence.

In practice, the forward algorithm works as follows. First you initialize the procedure by calculating for the states $j = 1, \ldots, N$ the quantities $\alpha_1(j) = Pr(z_1 = j)\omega_{j,y_1}$. Then you compute for the states $j = 1, \ldots, N$ the relationship $\alpha_t(j) = \sum_{i=1}^{N} \alpha_{t-1}(i)\gamma_{i,j}\omega_{j,y_t}$ for $t = 2, \ldots, T$.

Finally, you compute the marginal likelihood as $\Pr(y) = \sum_{j=1}^{N} \alpha_T(j)$. At each time t, we need to calculate N values of $\alpha_t(j)$, and each $\alpha_t(j)$ is a sum of N products of α_{t-1}, $\gamma_{i,j}$ and ω_{j,y_t}, hence TN^2 computations in total, which is much less than the $2TN^T$ calculations in the brute-force approach.

Going back to our example, we wish to calculate $\Pr(y_1 = 2, y_2 = 2, y_3 = 1)$. First we initialize and compute $\alpha_1(1)$ and $\alpha_1(2)$. We have:

$$\alpha_1(1) = \Pr(z_1 = 1)\omega_{1,y_1=2} = 1$$

because all animals are alive and captured in first winter. We also have:

$$\alpha_1(2) = \Pr(z_1 = 2)\omega_{2,y_1=2} = 0$$

Then we compute $\alpha_2(1)$ and $\alpha_2(2)$. We have:

$$\alpha_2(1) = \sum_{i=1}^{2} \alpha_1(i)\gamma_{i,1}\omega_{1,y_2=2}$$
$$= \gamma_{1,1}\omega_{1,y_2=2}$$
$$= \phi p$$

because $\alpha_1(2) = 0$. Also, we have:

$$\alpha_2(2) = \sum_{i=1}^{2} \alpha_1(i)\gamma_{i,2}\omega_{2,y_2=2}$$
$$= \gamma_{1,2}\omega_{2,y_2=2}$$
$$= (1 - \phi)0$$

Finally we compute $\alpha_3(1)$ and $\alpha_3(2)$. We have:

$$\alpha_3(1) = \sum_{i=1}^{2} \alpha_2(i)\gamma_{i,1}\omega_{1,y_3=1}$$
$$= \alpha_2(1)\gamma_{1,1}\omega_{1,y_3=1}$$
$$= \phi p\phi(1-p)$$

and:

$$\alpha_3(2) = \sum_{i=1}^{2} \alpha_2(i)\gamma_{i,2}\omega_{2,y_3=1}$$
$$= \alpha_2(1)\gamma_{1,2}\omega_{2,y_3=1}$$
$$= \phi p(1-\phi)1$$

Eventually, we compute $\Pr(y_1 = 2, y_2 = 2, y_3 = 1)$:

$$\Pr(y_1 = 2, y_2 = 2, y_3 = 1) = \alpha_3(1) + \alpha_3(2)$$
$$= \phi p(\phi)(1-p) + \phi p(1-\phi)$$
$$= \phi p(1 - \phi p)$$

You can check that we did in total $3 \times 2^2 = 12$ operations.

3.8.3 NIMBLE implementation

3.8.3.1 Do it yourself

In NIMBLE, we use functions to implement the forward algorithm. The only differences with the theory above is that i) we work on the log scale for numerical stability and ii) we use a matrix formulation of the recurrence.

First we write the density function:

```r
dHMMhomemade <- nimbleFunction(
  run = function(x = double(1),
                 probInit = double(1), # vector of initial states
                 probObs = double(2), # observation matrix
```

```
                     probTrans = double(2), # transition matrix
                     len = double(0, default = 0), # nb sampling occ
                     log = integer(0, default = 0)) {
  alpha <- probInit[1:2] # * probObs[1:2,x[1]] == 1 due to
                         # conditioning on first detection
  for (t in 2:len) {
  alpha[1:2]<-(alpha[1:2]%*%probTrans[1:2,1:2])*probObs[1:2,x[t]]
  }
  logL <- log(sum(alpha[1:2]))
  returnType(double(0))
  if (log) return(logL)
  return(exp(logL))
 }
)
```

In passing, this is the function you would maximize in a frequentist
approach (see Section 2.5). Then we write a function to simulate values
from a HMM:

```
rHMMhomemade <- nimbleFunction(
  run = function(n = integer(),
                 probInit = double(1),
                 probObs = double(2),
                 probTrans = double(2),
                 len = double(0, default = 0)) {
  returnType(double(1))
  z <- numeric(len)
  # all individuals alive at t = 0
  z[1] <- rcat(n = 1, prob = probInit[1:2])
  y <- z
  y[1] <- 2 # all individuals are detected at t = 0
  for (t in 2:len){
    # state at t given state at t-1
    z[t] <- rcat(n = 1, prob = probTrans[z[t-1],1:2])
    # observation at t given state at t
    y[t] <- rcat(n = 1, prob = probObs[z[t],1:2])
```

```r
    }
    return(y)
  })
```

We assign these functions to the global R environment:

```r
assign('dHMMhomemade', dHMMhomemade, .GlobalEnv)
assign('rHMMhomemade', rHMMhomemade, .GlobalEnv)
```

Now we resume our workflow:

```r
# code
hmm.survival <- nimbleCode({
  phi ~ dunif(0, 1) # prior survival
  p ~ dunif(0, 1) # prior detection
  # likelihood
  delta[1] <- 1           # Pr(alive t = 1) = 1
  delta[2] <- 0           # Pr(dead t = 1) = 0
  gamma[1,1] <- phi       # Pr(alive t -> alive t+1)
  gamma[1,2] <- 1 - phi   # Pr(alive t -> dead t+1)
  gamma[2,1] <- 0         # Pr(dead t -> alive t+1)
  gamma[2,2] <- 1         # Pr(dead t -> dead t+1)
  omega[1,1] <- 1 - p     # Pr(alive t -> non-detected t)
  omega[1,2] <- p         # Pr(alive t -> detected t)
  omega[2,1] <- 1         # Pr(dead t -> non-detected t)
  omega[2,2] <- 0         # Pr(dead t -> detected t)
  for (i in 1:N){
    y[i,1:T] ~ dHMMhomemade(probInit = delta[1:2],
                            probObs = omega[1:2,1:2], # observation
                            probTrans = gamma[1:2,1:2], # transition
                            len = T) # nb of sampling occasions
  }
})
# constants
my.constants <- list(N = nrow(y), T = 5)
# data
my.data <- list(y = y + 1)
```

```r
# initial values - no need to specify values for z anymore
initial.values <- function() list(phi = runif(1,0,1),
                                   p = runif(1,0,1))
# parameters to save
parameters.to.save <- c("phi", "p")
# MCMC details
n.iter <- 5000
n.burnin <- 1000
n.chains <- 2
```

And run NIMBLE:

```r
start_time <- Sys.time()
mcmc.output <- nimbleMCMC(code = hmm.survival,
                          constants = my.constants,
                          data = my.data,
                          inits = initial.values,
                          monitors = parameters.to.save,
                          niter = n.iter,
                          nburnin = n.burnin,
                          nchains = n.chains)
## |-------------|-------------|-------------|-------------|
## |-------------------------------------------------------|
## |-------------|-------------|-------------|-------------|
## |-------------------------------------------------------|
end_time <- Sys.time()
end_time - start_time
## Time difference of 40.35 secs
```

The numerical summaries are similar to those we obtained with the complete likelihood, and effective samples sizes are larger denoting better mixing:

```r
MCMCsummary(mcmc.output, round = 2)
##      mean    sd 2.5%  50% 97.5% Rhat n.eff
```

```
## p    0.61 0.06 0.49 0.61   0.72     1   1211
## phi  0.76 0.04 0.67 0.76   0.84     1   1483
```

3.8.3.2 Do it with `nimbleEcology`

Writing NIMBLE functions is not easy. Fortunately, the NIMBLE folks got you covered. They developed the package `nimbleEcology` that implements some of the most popular ecological models with latent states.

We will use the function `dHMMo` which provides the distribution of a hidden Markov model with time-independent transition matrix and time-dependent observation matrix. Why time-dependent observation matrix? Because we need to tell NIMBLE that detection at first encounter is 1.

We load the package:

```
library(nimbleEcology)
```

The NIMBLE code is:

```r
hmm.survival <- nimbleCode({
  phi ~ dunif(0, 1) # prior survival
  p ~ dunif(0, 1) # prior detection
  # likelihood
  delta[1] <- 1            # Pr(alive t = 1) = 1
  delta[2] <- 0            # Pr(dead t = 1) = 0
  gamma[1,1] <- phi        # Pr(alive t -> alive t+1)
  gamma[1,2] <- 1 - phi    # Pr(alive t -> dead t+1)
  gamma[2,1] <- 0          # Pr(dead t -> alive t+1)
  gamma[2,2] <- 1          # Pr(dead t -> dead t+1)
  omega[1,1,1] <- 0        # Pr(alive first -> non-detected first)
  omega[1,2,1] <- 1        # Pr(alive first -> detected first)
  omega[2,1,1] <- 1        # Pr(dead first -> non-detected first)
  omega[2,2,1] <- 0        # Pr(dead first -> detected first)
  for (t in 2:5){
    omega[1,1,t] <- 1 - p  # Pr(alive t -> non-detected t)
```

```r
    omega[1,2,t] <- p          # Pr(alive t -> detected t)
    omega[2,1,t] <- 1          # Pr(dead t -> non-detected t)
    omega[2,2,t] <- 0          # Pr(dead t -> detected t)
  }
  for (i in 1:N){
    y[i,1:5] ~ dHMMo(init = delta[1:2], # initial state probs
                     probObs = omega[1:2,1:2,1:5], # obs matrix
                     probTrans = gamma[1:2,1:2], # trans matrix
                     len = 5, # nb of sampling occasions
                     checkRowSums = 0) # skip validity checks
  }
})
```

You may see that we no longer have the states in the code as we use the marginalized likelihood. The dHMMo takes several arguments, including init the vector of initial state probabilities, probObs the observation matrix, probTrans the transition matrix and len the number of sampling occasions.

Next steps are similar to the workflow we used before. The only difference is that we do not need to specify initial values for the latent states:

```r
# constants
my.constants <- list(N = nrow(y))
# data
my.data <- list(y = y + 1)
# initial values - no need to specify values for z anymore
initial.values <- function() list(phi = runif(1,0,1),
                                   p = runif(1,0,1))

# parameters to save
parameters.to.save <- c("phi", "p")
# MCMC details
n.iter <- 5000
n.burnin <- 1000
n.chains <- 2
```

Now we run NIMBLE:

```r
start_time <- Sys.time()
mcmc.output <- nimbleMCMC(code = hmm.survival,
                          constants = my.constants,
                          data = my.data,
                          inits = initial.values,
                          monitors = parameters.to.save,
                          niter = n.iter,
                          nburnin = n.burnin,
                          nchains = n.chains)
## |-------------------|-------------------|-------------------|-------------------|
## |-----------------------------------------------------------------------------|
## |-------------------|-------------------|-------------------|-------------------|
## |-----------------------------------------------------------------------------|
end_time <- Sys.time()
end_time - start_time
## Time difference of 44 secs
```

Now we display the numerical summaries of the posterior distributions:

```r
MCMCsummary(mcmc.output, round = 2)
##      mean   sd 2.5%  50% 97.5% Rhat n.eff
## p    0.61 0.06 0.49 0.61  0.72    1  1453
## phi  0.76 0.04 0.67 0.76  0.84    1  1359
```

The results are similar what we obtained previously with our home-made marginalized likelihood (Section 3.8.3.1), or with the full likelihood (3.7).

3.9   Pooled encounter histories

We can go one step further to make convergence even faster. As mentionned earlier in Section 3.6.4, the likelihood of an HMM fitted to capture-recapture data often involves individuals that share the same encounter histories. Instead of repeating the same calculations several times, the likelihood contribution that is shared by say x individuals is raised to the power x in the likelihood of the whole dataset, hence making the same operations only once. This idea is used in routine in capture-recapture software. For Bayesian software, however, to my knowledge this trick has only been implemented in NIMBLE [Turek et al., 2016]. I am grateful to Chloé Nater for pointing this out to me.

In this section, we amend the NIMBLE functions we wrote for marginalizing latent states in Section 3.8 to express the likelihood using pooled encounter histories. We use a vector `size` that contains the number of individuals with the same encounter history.

The density function is the function `dHMMhomemade` to which we add a `size` argument, and raise the individual likelihood to the power `size`, or multiply by `size` as we work on the log scale `log(sum(alpha[1:2])) * size`:

```r
dHMMpooled <- nimbleFunction(
  run = function(x = double(1),
                 probInit = double(1),
                 probObs = double(2),
                 probTrans = double(2),
                 len = double(0),
                 size = double(0),
                 log = integer(0, default = 0)) {
    alpha <- probInit[1:2]
    for (t in 2:len) {
    alpha[1:2]<-(alpha[1:2]%*%probTrans[1:2,1:2])*probObs[1:2,x[t]]
    }
    logL <- log(sum(alpha[1:2])) * size
    returnType(double(0))
    if (log) return(logL)
    return(exp(logL))
  }
)
```

The `rHMMhomemade` function is renamed `rHMMpooled` for compatibility but remains unchanged:

```r
rHMMpooled <- nimbleFunction(
  run = function(n = integer(),
                 probInit = double(1),
                 probObs = double(2),
                 probTrans = double(2),
                 len = double(0),
                 size = double(0)) {
    returnType(double(1))
    z <- numeric(len)
    # all individuals alive at t = 0
    z[1] <- rcat(n = 1, prob = probInit[1:2])
    y <- z
    y[1] <- 2 # all individuals are detected at t = 0
    for (t in 2:len){
      # state at t given state at t-1
      z[t] <- rcat(n = 1, prob = probTrans[z[t-1],1:2])
      # observation at t given state at t
      y[t] <- rcat(n = 1, prob = probObs[z[t],1:2])
    }
    return(y)
  })
```

We assign these two function to the global R environment so that we can use them:

```r
assign('dHMMpooled', dHMMpooled, .GlobalEnv)
assign('rHMMpooled', rHMMpooled, .GlobalEnv)
```

You can now plug your pooled HMM density function in your NIMBLE code:

```r
hmm.survival <- nimbleCode({
  phi ~ dunif(0, 1) # prior survival
```

```r
  p ~ dunif(0, 1) # prior detection
  # likelihood
  delta[1] <- 1            # Pr(alive t = 1) = 1
  delta[2] <- 0            # Pr(dead t = 1) = 0
  gamma[1,1] <- phi        # Pr(alive t -> alive t+1)
  gamma[1,2] <- 1 - phi    # Pr(alive t -> dead t+1)
  gamma[2,1] <- 0          # Pr(dead t -> alive t+1)
  gamma[2,2] <- 1          # Pr(dead t -> dead t+1)
  omega[1,1] <- 1 - p      # Pr(alive t -> non-detected t)
  omega[1,2] <- p          # Pr(alive t -> detected t)
  omega[2,1] <- 1          # Pr(dead t -> non-detected t)
  omega[2,2] <- 0          # Pr(dead t -> detected t)
  for (i in 1:N){
    y[i,1:T] ~ dHMMpooled(probInit = delta[1:2],
                          probObs = omega[1:2,1:2], # obs matrix
                          probTrans = gamma[1:2,1:2], # trans matrix
                          len = T, # nb of sampling occasions
                          size = size[i]) # number of individuals
                                    # with encounter history i
  }
})
```

Before running NIMBLE, we need to actually pool individuals with the same encounter history together:

```r
y_pooled <- y %>%
  as_tibble() %>%
  group_by_all() %>% # group
  summarise(size = n()) %>% # count
  relocate(size) %>% # put size in front
  arrange(-size) %>% # sort along size
  as.matrix()
y_pooled
##      size winter 1 winter 2 winter 3 winter 4 winter 5
## [1,]   21        1        0        0        0        0
## [2,]    8        1        1        0        0        0
## [3,]    8        1        1        1        1        0
```

```
##  [4,]    4       1       1       0       0       1
##  [5,]    4       1       1       1       0       0
##  [6,]    2       1       0       0       1       0
##  [7,]    2       1       0       1       1       0
##  [8,]    2       1       1       0       1       0
##  [9,]    1       1       0       1       0       0
## [10,]    1       1       0       1       0       1
## [11,]    1       1       0       1       1       1
## [12,]    1       1       1       0       1       1
## [13,]    1       1       1       1       0       1
## [14,]    1       1       1       1       1       1
```

For example, we have 21 individuals with encounter history (1, 0, 0, 0, 0).

Now you can resume the NIMBLE workflow:

```r
my.constants <- list(N = nrow(y_pooled),
                     T = 5,
                     size = y_pooled[,'size'])
my.data <- list(y = y_pooled[,-1] + 1) # delete size from dataset
initial.values <- function() list(phi = runif(1,0,1),
                                   p = runif(1,0,1))
parameters.to.save <- c("phi", "p")
n.iter <- 5000
n.burnin <- 1000
n.chains <- 2
start_time <- Sys.time()
mcmc.output <- nimbleMCMC(code = hmm.survival,
                          constants = my.constants,
                          data = my.data,
                          inits = initial.values,
                          monitors = parameters.to.save,
                          niter = n.iter,
                          nburnin = n.burnin,
                          nchains = n.chains)
```

```
## |------------------|------------------|-----------------|------------------|
## |------------------------------------------------------------------------|
## |------------------|------------------|-----------------|------------------|
## |------------------------------------------------------------------------|
end_time <- Sys.time()
end_time - start_time
## Time difference of 36.05 secs
MCMCsummary(mcmc.output, round = 2)
##      mean   sd 2.5%  50% 97.5% Rhat n.eff
## p    0.61 0.06 0.49 0.61  0.72    1  1455
## phi  0.76 0.04 0.67 0.76  0.84    1  1524
```

The results are the same as those obtained previously. The gain in computation times will be bigger for more complex models.

The pooled likelihood is not yet implemented in `nimbleEcology`, but you can hack the code for the function `dHMMo` https://github.com/nimble-dev/nimbleEcology/blob/master/R/dHMM.R to implement it yourself by adding a `size` argument.

3.10 Decoding after marginalization

If you need to infer the latent states, and you cannot afford the computation times of the complete likelihood of Section 3.6.4, you can still use the marginal likelihood with the forward algorithm of Section 3.8.2. You will need an extra step to decode the latent states with the Viterbi algorithm. The Viterbi algorithm allows you to compute the sequence of states that is most likely to have generated the sequence of observations.

3.10.1 Theory

In our simulated dataset, animal #15 has the encounter history (2, 1, 1, 1, 1) which was generated from the sequence of states (1, 1, 2, 2, 2) with survival probability $\phi = 0.8$ and detection probability $p = 0.6$.

Imagine you do not know the truth. What is the chance that animal #15 was alive throughout the study when observing the encounter history detected in first winter, then missed in each subsequent winter? The chance of being alive in the first winter and detected when alive is 1. The chance of being alive in the second winter and non-detected is $0.8 \times (1 - 0.6) = 0.32$. The same goes for third, fourth and fifth winters. In total, the probability of being alive throughout the study for an animal with encounter history $(2, 1, 1, 1, 1)$ is $1 \times 0.32 \times 0.32 \times 0.32 \times 0.32 = 0.01048576$.

Now what is the chance that animal #15 was alive, then dead for the rest of the study, when observing the encounter history $(2, 1, 1, 1, 1)$? And the chance of being alive in first and second winters, then dead after when observing the same encounter history? And so on. You need to enumerate all possible sequences of states and compute the probability for each of them, and choose the most probable sequence, that is with maximum probability. In our example, we would need to compute $2^5 = 32$ of these probabilities, and N^T in general. Needless to say, these calculations quickly become cumbersome, if not impossible, as the number of states and/or the number of sampling occasions increases.

This is where the Viterbi algorithm comes in. The idea is to decompose this overall complex problem in a sequence of smaller problems that are easier to solve. If dynamic programming rings a bell, the Viterbi algorithm should look familiar to you. The Viterbi algorithm is based on the fact that the optimal path to each winter and each state can be deduced from the optimal path to the previous winter and each state.

For first winter, the probability of being alive and detected is 1, while the probability of being dead and detected is 0. Now what is the probability of being alive in the second winter and non-detected? If the animal was alive in the first winter, it remains alive and is missed with probability $1 \times \phi(1 - p) = 0.32$. If it was dead in the first winter, then this probability is 0. The maximum probability is 0.32 obviously so the most probable scenario to being alive in the second winter is being alive in the first winter. What about being dead in the second winter? If the animal was alive in first winter, then the probability is $1 \times (1 - \phi) \times 1 = 0.2$. If dead, then this probability is $0 \times 0 \times (1 - p) = 0$. The maximum probability is 0.2 obviously so the most probable scenario to being dead in the second winter is being alive in the first winter. Doing these calculations for third, fourth and fifth winters, we get the probabilities:

winter	1	2	3	4	5
state alive	1	0.32 = max(0, 0.32)	0.2304 = max(0, 0.2304)	0.110592 = max(0, 0.110592)	0.053084416 = max(0, 0.05308416
state dead	0	1 = max(1, 0.2)	1 = max(0.096, 1)	1 = max(0.04608, 1)	1 = max(0.0221184, 1)
obs.	det.	missed	missed	missed	missed

Finally, to get (or decode), the optimal path, you work backwards and trace back the previous value that yielded the maximum probability. The most probable state in the last winter is dead (1 > 0.05308416), dead again in the fourth winter (1 > 0.110592), dead in the third winter (1 > 0.2304), alive in the second winter (1 > 0.48) and alive in the first winter (1 > 0). According to the Viterbi algorithm, the sequence of states that most likely generated the sequence of observations (2, 1, 1, 1, 1) is alive then dead, dead, dead and dead or (1 2 2 2 2). This differs slightly from the actual sequence of states (1, 1, 2, 2, 2) in that the state in second winter is decoded dead while animal #15 only dies in third winter.

In contrast to the brute force approach, calculations are not duplicated but stored and used again like in the forward algorithm. Briefly speaking, the Viterbi algorithm works like the forward algorithm where sums are replaced by calculating maximums.

In practice, the Viterbi algorithm works as illustrated in Figure 3.1. First you initialize the procedure by calculating at $t = 1$ for all states $j = 1, \ldots, N$ the values $\nu_1(j) = Pr(z_1 = j)\omega_{j,y_1}$. Then you compute for all states $j = 1, \ldots, N$ the values $\nu_t(j) = \max_{i=1,\ldots,N} \nu_{t-1}(i)\gamma_{i,j}\omega_{j,y_t}$ for $t = 2, \ldots, T$. Here at each time t we determine the probability of the best path ending at each of the states $j = 1, \ldots, N$. Finally, you compute the probability of the best global path $\max_{j=1,\ldots,N} \nu_T(j)$.

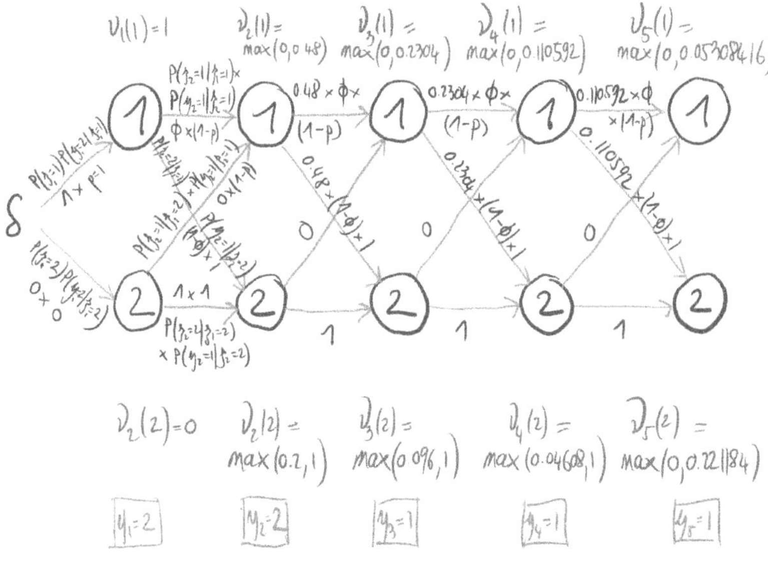

FIGURE 3.1: Graphical representation of the Viterbi algorithm with $\phi = 0.8$ and $p = 0.6$. States are alive $z = 1$ or dead $z = 2$ and observations are non-detected $y = 1$ or detected $y = 2$.

3.10.2   Implementation

Let's write an R function to implement the Viterbi algorithm. As parameters, our function will take the transition and observation matrices, the vector of initial state probabilities and the observed sequence of detections and non-detections for which you aim to compute the sequence of states from which it was most likely generated:

```r
# getViterbi() returns sequence of states that most
# likely generated sequence of observations
# adapted from https://github.com/vbehnam/viterbi
getViterbi <- function(Omega, Gamma, delta, y) {
# Omega: transition matrix
# Gamma: observation matrix
# delta: vector of initial state probabilities
# y: observed sequence of detections and non-detections
```

```r
# get number of states and sampling occasions
N <- nrow(Gamma)
T <- length(y)

# nu is the corresponding likelihood
nu <- matrix(0, nrow = N, ncol = T)
# zz contains the most likely states up until this point
zz <- matrix(0, nrow = N, ncol = T)
firstObs <- y[1]

# fill in first columns of both matrices
#nu[,1] <- initial * emission[,firstObs]
#zz[,1] <- 0
nu[,1] <- c(1,0) # initial = (1, 0) * emission[,firstObs] = (1, 0)
zz[,1] <- 1 # alive at first occasion

for (i in 2:T) {
    for (j in 1:N) {
      obs <- y[i]
      # initialize to -1, then overwritten by
      # for loop coz all possible values are >= 0
      nu[j,i] <- -1
      # loop to find max and argmax for k
      for (k in 1:N) {
        value <- nu[k,i-1] * Gamma[k,j] * Omega[j,obs]
        if (value > nu[j,i]) {
          # maximizing for k
          nu[j,i] <- value
          # argmaximizing for k
          zz[j,i] <- k
        }
      }
    }
  }
  # mlp = most likely path
```

```r
  mlp <- numeric(T)
  # argmax for stateSeq[,T]
  am <- which.max(nu[,T])
  mlp[T] <- zz[am,T]


  # backtrace using backpointers
  for (i in T:2) {
    zm <- which.max(nu[,i])
    mlp[i-1] <- zz[zm,i]
  }
  return(mlp)
}
```

Note that instead of writing your own R function, you could use a built-in function from an existing R package to implement the Viterbi algorithm (for example, the `viterbi()` function from the HMM and `depmixS4` packages), and call it from NIMBLE as we have seen in Section 2.4.2. The difficulty is that HMM for capture-recapture data have specific features that make standard functions not adapted and requires coding your own Viterbi function. In particular, we have to deal with detection at first encounter, which is not estimated but is always one because an individual has to be captured to be marked and released for the first time. Also, our transition and observation matrices are not always homogeneous and may depend on time.

Let's test our `getViterbi()` function with our previous example. Remember animal #15 has the encounter history (2, 1, 1, 1, 1) which was generated from the sequence of states (1, 1, 2, 2, 2). Applying our function to that animal encounter history, we get:

```r
delta # Vector of initial states probabilities
## [1] 1 0
Gamma # Transition matrix
##       [,1] [,2]
## [1,]  0.8  0.2
## [2,]  0.0  1.0
```

```
Omega # Observation matrix
##       [,1] [,2]
## [1,]  0.4  0.6
## [2,]  1.0  0.0
getViterbi(Omega = Omega,
           Gamma = Gamma,
           delta = delta,
           y = y[15,] + 1)
## [1] 1 2 2 2 2
```

The Viterbi algorithm does pretty well at recovering the latent states,
despite incorrectly decoding a death in the second winter while individual #15 only dies in the third winter. We obtained the same results
by implementing the Viterbi algorithm by hand in Section 3.10.1.

Now that we have a function that implements the Viterbi algorithm, we
can use it with our MCMC outputs. You have two options, either you
apply Viterbi to each MCMC iteration then you compute the posterior
median or mode path for each individual, or you compute the posterior
mean or median of the transition and observation matrices then you
apply Viterbi to each individual encounter history.

For both options, we will need the values from the posterior distributions of survival and detection probabilities:

```
phi <- c(mcmc.output$chain1[,'phi'], mcmc.output$chain2[,'phi'])
p <- c(mcmc.output$chain1[,'p'], mcmc.output$chain2[,'p'])
```

3.10.3   Compute first, average after

First option is to apply Viterbi to each MCMC sample, then to compute
median of the MCMC Viterbi paths for each observed sequence:

```
niter <- length(p)
T <- 5
res <- matrix(NA, nrow = nrow(y), ncol = T)
```

```r
for (i in 1:nrow(y)){
  res_mcmc <- matrix(NA, nrow = niter, ncol = T)
  for (j in 1:niter){
    # Initial states
    delta <- c(1, 0)
    # Transition matrix
    transition <- matrix(NA, 2, 2)
    transition[1,1] <- phi[j]       # Pr(alive t -> alive t+1)
    transition[1,2] <- 1 - phi[j]   # Pr(alive t -> dead t+1)
    transition[2,1] <- 0            # Pr(dead t -> alive t+1)
    transition[2,2] <- 1            # Pr(dead t -> dead t+1)
    # Observation matrix
    emission <- matrix(NA, 2, 2)
    emission[1,1] <- 1 - p[j]       # Pr(alive t -> non-detected t)
    emission[1,2] <- p[j]           # Pr(alive t -> detected t)
    emission[2,1] <- 1              # Pr(dead t -> non-detected t)
    emission[2,2] <- 0              # Pr(dead t -> detected t)
    res_mcmc[j,1:T] <- getViterbi(Omega = emission,
                                  Gamma = transition,
                                  delta = delta,
                                  y = y[i,] + 1)
  }
  res[i, 1:length(y[1,])] <- apply(res_mcmc, 2, median)
}
```

You can compare the Viterbi decoding to the actual states z:

Decoding is correct except that the alive actual state is often decoded as the dead state by the Viterbi algorithm. Note that here we compute the Viterbi paths after we run NIMBLE. You could turn the R function `getViterbi()` into a NIMBLE function and plug it in your model code to apply Viterbi. This would not make any difference except perhaps to increase MCMC computation times.

3.10.4   Average first, compute after

Second option is to compute the posterior mean of the observation and transition matrices, then to apply Viterbi:

```r
# Initial states
delta <- c(1, 0)
# Transition matrix
transition <- matrix(NA, 2, 2)
transition[1,1] <- mean(phi)        # Pr(alive t -> alive t+1)
transition[1,2] <- 1 - mean(phi)    # Pr(alive t -> dead t+1)
transition[2,1] <- 0                # Pr(dead t -> alive t+1)
transition[2,2] <- 1                # Pr(dead t -> dead t+1)
# Observation matrix
emission <- matrix(NA, 2, 2)
emission[1,1] <- 1 - mean(p)        # Pr(alive t -> non-detected t)
emission[1,2] <- mean(p)            # Pr(alive t -> detected t)
emission[2,1] <- 1                  # Pr(dead t -> non-detected t)
emission[2,2] <- 0                  # Pr(dead t -> detected t)
res <- matrix(NA, nrow = nrow(y), ncol = T)
for (i in 1:nrow(y)){
  res[i, 1:length(y[1,]) ] <- getViterbi(Omega = emission,
                                          Gamma = transition,
                                          delta = delta,
                                          y = y[i,] + 1)
}
```

Again, you can compare the result of the Viterbi decoding to the actual states we simulated and used to generate the observations:

The results are very similar to those we obtained in Section 3.10.3, and Figure 3.3 is indistinguishable from Figure 3.2.

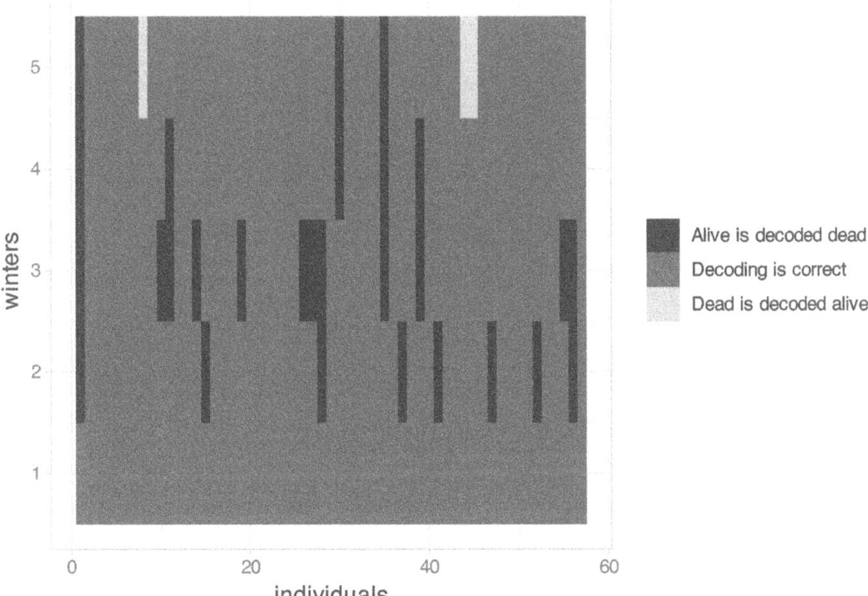

FIGURE 3.2: Comparison of the actual sequences of states to the sequences of states decoded with Viterbi and the – average first, compute after – method.

3.11   Summary

- A HMM is a model that consists of two parts: i) an unobserved sequence of discrete random variables – the states – satisfying the Markovian property (future states depends on current states only and not on past states) and ii) an observed sequence of discrete random variables – the observations – depending only on the current state.

- The Bayesian approach together with MCMC simulations allow estimating survival and detection probabilities as well as individual latent states alive or dead with the complete likelihood. If you can afford the computation times, then using the complete likelihood is the easiest path for model fitting.

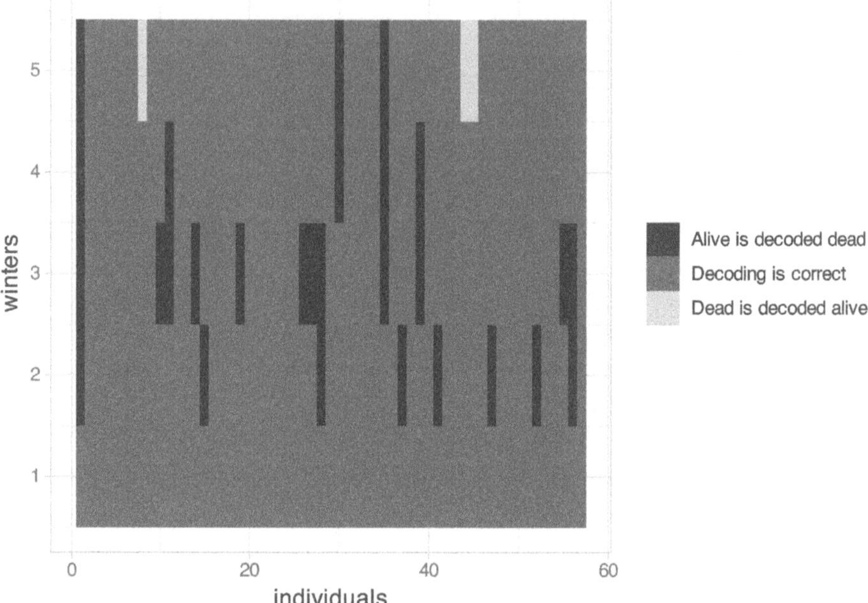

FIGURE 3.3: Comparison of the actual sequences of states to the sequences of states decoded with Viterbi and the – compute first, average after method.

- If you do not need to infer the latent states, you can use the marginal likelihood via the forward algorithm. By avoiding to sample the latent states, you usually get better mixing and faster convergence.

- If you do need to infer the latent states, and you cannot afford the computation times of the complete likelihood, then you can go for the marginal likelihood in conjunction with the Viterbi algorithm to decode the latent states.

- If the computational burden is still an issue, and you have individuals that share the same encounter history, you can use a pooled likelihood to speed up the marginal likelihood evaluation and MCMC convergence.

3.12 Suggested reading

- A landmark paper on HMM is Rabiner [1989].

- Check out Jurafsky and Martin [2023] for a nice introduction to HMM and Zucchini et al. [2016] for an excellent book that covers theory and applications.

- The paper by McClintock et al. [2020] reviews the applications of HMM in ecology.

- The package `nimbleEcology` is developed by Goldstein et al. [2021]; see Ponisio et al. [2020] for its application to occupancy and N−mixture models.

Part II

Transitions

This second part `Transitions` will teach you all about capture-recapture models for open populations, with reproducible R code to ease the learning process. The code and data are available at https://github.com /oliviergimenez/banana-book/tree/master/appendix.

4

Alive and dead

4.1 Introduction

In this fourth chapter, you will learn about the Cormack-Jolly-Seber model that allows estimating survival based on capture-recapture data. You will also see how to deal with covariates to try and explain temporal and/or individual variation in survival. This chapter will also be the opportunity to introduce tools to compare models and assess their quality of fit to data.

4.2 The Cormack-Jolly-Seber (CJS) model

In Chapter 3, we introduced a capture-recapture model with constant survival and detection probabilities which we formulated as a HMM and fitted to data in NIMBLE. Historically, however, it was a slightly more complicated model that was first proposed – the so-called Cormack-Jolly-Seber (CJS) model – in which survival and recapture probabilities are time-varying. This feature of the CJS model is useful to account for variation due to environmental conditions in survival or to sampling effort in detection. Schematically the CJS model can be represented this way:

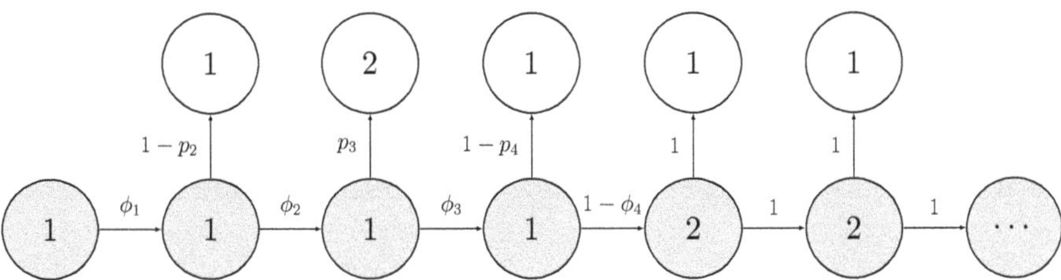

DOI: 10.1201/9781003244097-4

Note that the states (in gray) and the observations (in white) do not change. We still have $z = 1$ for alive, $z = 2$ for dead, $y = 1$ for non-detected, and $y = 2$ for detected.

Parameters are now indexed by time. The survival probability is defined as the probability of staying alive (or "ah, ha, ha, ha, stayin' alive" like the Bee Gees would say) over the interval between t and $t + 1$, that is $\phi_t = \Pr(z_{t+1} = 1|z_t = 1)$. The detection probability is defined as the probability of being observed at t given you're alive at t, that is $p_t = \Pr(y_t = 1|z_t = 1)$. It is important to bear in mind for later (see Section 4.8) that survival operates over an interval while detection occurs at a specific time.

The CJS model is named after the three statisticians – Richard Cormack, George Jolly and George Seber – who each published independently a paper introducing more or less the same approach, a year apart! In fact, Richard Cormack and George Jolly were working in the same corridor in Scotland back in the 1960s. They would meet every day at coffee and would play some game together, but never mention work and were not aware of each other's work.

4.3   Capture-recapture data

Before we turn to fitting the CJS model to actual data, let's talk about capture-recapture for a minute. We said in Section 3.6.1 that animals are individually marked. This can be accomplished in two ways, either with artificial marks like rings for birds or ear tags for mammals, or (non-invasive) natural marks like coat patterns or feces DNA sequencing (Figure 4.1).

Throughout this chapter, we will use data on the White-throated Dipper (*Cinclus cinclus*; dipper hereafter) kindly provided by Gilbert Marzolin (Figure 4.2). In total, 294 dippers with known sex and wing length were captured and recaptured between 1981 and 1987 during the March-June period. Birds were at least 1 year old when initially banded.

FIGURE 4.1: Animal individual marking. Top-left: rings (credit: Emannuelle Cam); Top-right: ear-tags (credit: Jessica Cachelou); Bottom left: coat patterns (credit: Fridolin Zimmermann); Bottom right: ADN feces (credit: Alexander Kopatz)

The data look like:

```
dipper <- read_csv("dipper.csv")
dipper
```

```
## # A tibble: 294 x 9
##    year_1981 year_1982 year_1983 year_1984 year_1985
##        <dbl>     <dbl>     <dbl>     <dbl>     <dbl>
## 1         1         1         1         1         1
## 2         1         1         1         1         1
## 3         1         1         1         1         0
## 4         1         1         1         1         0
## 5         1         1         0         1         1
## 6         1         1         0         0         0
## 7         1         1         0         0         0
```

FIGURE 4.2: White-throated Dipper (*Cinclus cinclus*). Credit: Gilbert
Marzolin.

```
## 8           1           1           0           0           0
## 9           1           1           0           0           0
## 10          1           1           0           0           0
## # i 284 more rows
## # i 4 more variables: year_1986 <dbl>, year_1987 <dbl>,
## #   sex <chr>, wing_length <dbl>
```

The first seven columns are years in which Gilbert went on the field
and captured the birds. A 0 stands for a non-detection, and a 1 for a
detection. The eighth column informs on the sex of the bird, with F for
female and M for male. The last column gives a measure wing length
the first time a bird was captured.

4.4   Fitting the CJS model to the dipper data with NIMBLE

To write the NIMBLE code corresponding to the CJS model, we only
need to make a few adjustments to the NIMBLE code for the model with
constant parameters from Section 3.7. The main modification concerns

the observation and transition matrices which we need to make time-varying. These matrices therefore become arrays and inherit a third dimension for time, besides that for rows and columns. Also we need priors for all time-varying survival `phi[t] ~ dunif(0, 1)` and detection `p[t] ~ dunif(0, 1)` probabilities. We write:

```
...
# parameters
  delta[1] <- 1                       # Pr(alive t = 1) = 1
  delta[2] <- 0                       # Pr(dead t = 1) = 0
  for (t in 1:(T-1)){
    phi[t] ~ dunif(0, 1)              # prior survival
    gamma[1,1,t] <- phi[t]            # Pr(alive t -> alive t+1)
    gamma[1,2,t] <- 1 - phi[t]        # Pr(alive t -> dead t+1)
    gamma[2,1,t] <- 0                 # Pr(dead t -> alive t+1)
    gamma[2,2,t] <- 1                 # Pr(dead t -> dead t+1)
    p[t] ~ dunif(0, 1)               # prior detection
    omega[1,1,t] <- 1 - p[t]          # Pr(alive t -> non-detected t)
    omega[1,2,t] <- p[t]              # Pr(alive t -> detected t)
    omega[2,1,t] <- 1                 # Pr(dead t -> non-detected t)
    omega[2,2,t] <- 0                 # Pr(dead t -> detected t)
  }
...
```

The likelihood does not change, except that the time-varying observation and transition matrices need to be used appropriately. Also, because we now deal with several cohorts of animals first captured, marked and released each year (in contrast with a single cohort as in Chapter 3), we need to start the loop over time from the first capture for each individual. Therefore, we write:

```
...
# likelihood
  for (i in 1:N){
    z[i,first[i]] ~ dcat(delta[1:2])
    for (j in (first[i]+1):T){
      z[i,j] ~ dcat(gamma[z[i,j-1], 1:2, j-1])
```

```
      y[i,j] ~ dcat(omega[z[i,j], 1:2, j-1])
    }
  }
```
...

At first capture, all individuals are alive `z[i,first[i]] ~ dcat(delta[1:2])` and detection is 1, then after first capture `for (j in (first[i]+1):T)` we apply the transition and observation matrices. Overall, the code looks like:

```
hmm.phitpt <- nimbleCode({
  # parameters
  delta[1] <- 1                    # Pr(alive t = 1) = 1
  delta[2] <- 0                    # Pr(dead t = 1) = 0
  for (t in 1:(T-1)){
    phi[t] ~ dunif(0, 1)           # prior survival
    gamma[1,1,t] <- phi[t]         # Pr(alive t -> alive t+1)
    gamma[1,2,t] <- 1 - phi[t]     # Pr(alive t -> dead t+1)
    gamma[2,1,t] <- 0              # Pr(dead t -> alive t+1)
    gamma[2,2,t] <- 1              # Pr(dead t -> dead t+1)
    p[t] ~ dunif(0, 1)             # prior detection
    omega[1,1,t] <- 1 - p[t]       # Pr(alive t -> non-detected t)
    omega[1,2,t] <- p[t]           # Pr(alive t -> detected t)
    omega[2,1,t] <- 1             # Pr(dead t -> non-detected t)
    omega[2,2,t] <- 0             # Pr(dead t -> detected t)
  }
  # likelihood
  for (i in 1:N){
    z[i,first[i]] ~ dcat(delta[1:2])
    for (j in (first[i]+1):T){
      z[i,j] ~ dcat(gamma[z[i,j-1], 1:2, j-1])
      y[i,j] ~ dcat(omega[z[i,j], 1:2, j-1])
    }
  }
})
```

We extract the detections and non-detections from the data:

```r
y <- dipper %>%
  select(year_1981:year_1987) %>%
  as.matrix()
```

We get the occasion of first capture for all individuals, by finding the position of detections in the encounter history (`which(x !=0)`), and keeping the first one:

```r
first <- apply(y, 1, function(x) min(which(x !=0)))
```

Now we specify the constants:

```r
my.constants <- list(N = nrow(y),     # number of animals
                     T = ncol(y),     # number of sampling occasions
                     first = first) # first capture for all animales
```

We then put the data in a list. We add 1 to the data to code non-detections as 1's detections as 2's (see Section 3.7).

```r
my.data <- list(y = y + 1)
```

Let's write a function for the initial values. For the latent states, we go the easy way, and say that all individuals are alive through the study period:

```r
zinits <- y + 1 # non-detection -> alive
zinits[zinits == 2] <- 1 # dead -> alive
initial.values <- function() list(phi = runif(my.constants$T-1,0,1),
                                   p = runif(my.constants$T-1,0,1),
                                   z = zinits)
```

We specify the parameters we would like to monitor, survival and detection probabilities here:

```
parameters.to.save <- c("phi", "p")
```

We provide MCMC details:

```
n.iter <- 5000
n.burnin <- 1000
n.chains <- 2
```

And we run NIMBLE:

```
mcmc.phitpt <- nimbleMCMC(code = hmm.phitpt,
                          constants = my.constants,
                          data = my.data,
                          inits = initial.values,
                          monitors = parameters.to.save,
                          niter = n.iter,
                          nburnin = n.burnin,
                          nchains = n.chains)
```

We may have a look to the numerical summaries:

```
MCMCsummary(mcmc.phitpt, params = c("phi","p"), round = 2)
##          mean   sd 2.5%  50% 97.5% Rhat n.eff
## phi[1] 0.72 0.13 0.45 0.72  0.96 1.00   553
## phi[2] 0.45 0.07 0.32 0.45  0.60 1.01  1175
## phi[3] 0.48 0.06 0.36 0.48  0.60 1.00  1320
## phi[4] 0.63 0.06 0.51 0.63  0.75 1.01  1041
## phi[5] 0.60 0.06 0.49 0.60  0.71 1.01  1030
## phi[6] 0.70 0.14 0.49 0.69  0.97 1.00    86
## p[1]   0.67 0.13 0.39 0.67  0.90 1.00   923
## p[2]   0.87 0.08 0.68 0.88  0.98 1.00   690
## p[3]   0.88 0.06 0.73 0.89  0.97 1.00   840
## p[4]   0.88 0.06 0.75 0.89  0.96 1.02   860
## p[5]   0.90 0.05 0.79 0.91  0.98 1.01   740
## p[6]   0.76 0.14 0.51 0.77  0.98 1.00    95
```

There is not so much time variation in the detection probability which is estimated high around 0.90. Note that p[1] corresponds to the detection probability in 1982 that is p_2, p[2] to detection in 1983 therefore p_3, and so on. The dippers seem to have experienced a decrease in survival in years 1982–1983 (phi[2]) and 1983–1984 (phi[4]). We will get back to that in Section 4.8.

You may have noticed the small effective sample size for the last survival (phi[6]) and detection (p[6]) probabilities. Let's have a look to mixing for parameter phi[6] for example:

```
priors <- runif(3000, 0, 1)
MCMCtrace(object = mcmc.phitpt,
          ISB = FALSE,
          exact = TRUE,
          params = c("phi[6]"),
          pdf = FALSE,
          priors = priors)
```

Trace - phi[6]   **Density - phi[6]**

Value: 1.0, 0.8, 0.6, 0.4 — Iteration: 0, 1000, 3000

Density: 3.0, 2.0, 1.0, 0.0 — 54.4% overlap — Parameter estimate: 0.4, 0.6, 0.8, 1.0

Clearly mixing (left panel in the plot above) is bad and there is a big overlap between the prior and the posterior for this parameter (right panel) suggesting that its prior was not well updated with the data. What is going on? If you could inspect the likelihood of the CJS model, you would realize that these two parameters ϕ_6 and p_7 appear only as the product $\phi_6 p_7$ and cannot be estimated separately. In other words, one of these parameters is redundant, and you'd need an extra sampling occasion to be able to disentangle them. This is not a big issue as long

as you're aware of it and you do not attempt to ecologically interpret
these parameters.

4.5   CJS model derivatives

Besides the model we considered with constant parameters (see Chapter 3) and the CJS model with time-varying parameters, you might want
to fit in-between or models with time variation on either detection or
survival.

But I realize that we did not actually fit the model with constant parameters from Chapter 3. Let's do it. You should be familiar with the
process by now:

```r
# NIMBLE code
hmm.phip <- nimbleCode({
  phi ~ dunif(0, 1)        # prior survival
  p ~ dunif(0, 1)          # prior detection
  # likelihood
  gamma[1,1] <- phi        # Pr(alive t -> alive t+1)
  gamma[1,2] <- 1 - phi    # Pr(alive t -> dead t+1)
  gamma[2,1] <- 0          # Pr(dead t -> alive t+1)
  gamma[2,2] <- 1          # Pr(dead t -> dead t+1)
  delta[1] <- 1            # Pr(alive t = 1) = 1
  delta[2] <- 0            # Pr(dead t = 1) = 0
  omega[1,1] <- 1 - p      # Pr(alive t -> non-detected t)
  omega[1,2] <- p          # Pr(alive t -> detected t)
  omega[2,1] <- 1          # Pr(dead t -> non-detected t)
  omega[2,2] <- 0          # Pr(dead t -> detected t)
  for (i in 1:N){
    z[i,first[i]] ~ dcat(delta[1:2])
    for (j in (first[i]+1):T){
      z[i,j] ~ dcat(gamma[z[i,j-1], 1:2])
      y[i,j] ~ dcat(omega[z[i,j], 1:2])
    }
```

```r
  }
})
# occasions of first capture
first <- apply(y, 1, function(x) min(which(x !=0)))
# constants
my.constants <- list(N = nrow(y),
                     T = ncol(y),
                     first = first)
# data
my.data <- list(y = y + 1)
# initial values
zinits <- y + 1 # non-detection -> alive
zinits[zinits == 2] <- 1 # dead -> alive
initial.values <- function() list(phi = runif(1,0,1),
                                  p = runif(1,0,1),
                                  z = zinits)
# parameters to monitor
parameters.to.save <- c("phi", "p")
# MCMC details
n.iter <- 5000
n.burnin <- 1000
n.chains <- 2
# run NIMBLE
mcmc.phip <- nimbleMCMC(code = hmm.phip,
                        constants = my.constants,
                        data = my.data,
                        inits = initial.values,
                        monitors = parameters.to.save,
                        niter = n.iter,
                        nburnin = n.burnin,
                        nchains = n.chains)
## |-------------|-------------|-------------|-------------|
## |-------------------------------------------------------|
## |-------------|-------------|-------------|-------------|
## |-------------------------------------------------------|
# numerical summaries
```

```
MCMCsummary(mcmc.phip, round = 2)
##      mean   sd 2.5%  50% 97.5% Rhat n.eff
## p    0.90 0.03 0.83 0.90  0.94 1.00   661
## phi  0.56 0.02 0.51 0.56  0.61 1.01  1633
```

Let's now turn to the model with time-varying survival and constant detection. We modify the CJS model NIMBLE code by no longer having the observation matrix time-specific. I'm just providing the model code to save space:

```
hmm.phitp <- nimbleCode({
  for (t in 1:(T-1)){
    phi[t] ~ dunif(0, 1)        # prior survival
    gamma[1,1,t] <- phi[t]      # Pr(alive t -> alive t+1)
    gamma[1,2,t] <- 1 - phi[t]  # Pr(alive t -> dead t+1)
    gamma[2,1,t] <- 0           # Pr(dead t -> alive t+1)
    gamma[2,2,t] <- 1           # Pr(dead t -> dead t+1)
  }
  p ~ dunif(0, 1) # prior detection
  delta[1] <- 1             # Pr(alive t = 1) = 1
  delta[2] <- 0             # Pr(dead t = 1) = 0
  omega[1,1] <- 1 - p      # Pr(alive t -> non-detected t)
  omega[1,2] <- p          # Pr(alive t -> detected t)
  omega[2,1] <- 1          # Pr(dead t -> non-detected t)
  omega[2,2] <- 0          # Pr(dead t -> detected t)
  # likelihood
  for (i in 1:N){
    z[i,first[i]] ~ dcat(delta[1:2])
    for (j in (first[i]+1):T){
      z[i,j] ~ dcat(gamma[z[i,j-1], 1:2, j-1])
      y[i,j] ~ dcat(omega[z[i,j], 1:2])
    }
  }
})
```

We obtain the following numerical summaries for parameters, confirming high detection and temporal variation in survival:

```
##           mean   sd 2.5%  50% 97.5% Rhat n.eff
## phi[1] 0.63 0.11 0.41 0.63  0.84    1  1407
## phi[2] 0.46 0.07 0.33 0.46  0.60    1  1350
## phi[3] 0.48 0.06 0.37 0.48  0.59    1  1586
## phi[4] 0.62 0.06 0.51 0.62  0.73    1  1528
## phi[5] 0.61 0.06 0.50 0.61  0.71    1  1463
## phi[6] 0.59 0.06 0.48 0.59  0.71    1  1101
## p      0.89 0.03 0.82 0.89  0.94    1   595
```

Now the model with time-varying detection and constant survival, for which the NIMBLE code has a constant over time transition matrix:

```r
hmm.phipt <- nimbleCode({
  phi ~ dunif(0, 1)           # prior survival
  gamma[1,1] <- phi           # Pr(alive t -> alive t+1)
  gamma[1,2] <- 1 - phi       # Pr(alive t -> dead t+1)
  gamma[2,1] <- 0             # Pr(dead t -> alive t+1)
  gamma[2,2] <- 1             # Pr(dead t -> dead t+1)
  delta[1] <- 1               # Pr(alive t = 1) = 1
  delta[2] <- 0               # Pr(dead t = 1) = 0
  for (t in 1:(T-1)){
    p[t] ~ dunif(0, 1)            # prior detection
    omega[1,1,t] <- 1 - p[t] # Pr(alive t -> non-detected t)
    omega[1,2,t] <- p[t]     # Pr(alive t -> detected t)
    omega[2,1,t] <- 1        # Pr(dead t -> non-detected t)
    omega[2,2,t] <- 0        # Pr(dead t -> detected t)
  }
  # likelihood
  for (i in 1:N){
    z[i,first[i]] ~ dcat(delta[1:2])
    for (j in (first[i]+1):T){
      z[i,j] ~ dcat(gamma[z[i,j-1], 1:2])
```

```
    y[i,j] ~ dcat(omega[z[i,j], 1:2, j-1])
  }
 }
})
```

Numerical summaries for the parameters are:

```
##         mean   sd 2.5%  50% 97.5% Rhat n.eff
## phi    0.56 0.03 0.51 0.56  0.61 1.00  1296
## p[1]   0.74 0.12 0.49 0.74  0.93 1.00  1369
## p[2]   0.84 0.08 0.66 0.85  0.98 1.01   744
## p[3]   0.84 0.07 0.68 0.85  0.96 1.00   722
## p[4]   0.89 0.05 0.77 0.89  0.97 1.00   962
## p[5]   0.92 0.04 0.81 0.92  0.98 1.00  1123
## p[6]   0.90 0.07 0.74 0.91  1.00 1.01   378
```

We note that these two models do no longer have parameter redundancy issues.

Before we move to the next section, you might ask how to fit these models with `nimbleEcology` as in Section 3.8.3. Conveniently there are specific NIMBLE functions for the marginalized likelihood of the CJS model and its derivatives. These function are named generically `dCJSxx()` where the first x is for survival and the second x is for detection and x can be s for scalar or constant or v for vector or time-dependent. For example, to implement the model with constant survival and detection probabilities, you would use `dCJSss()`:

```r
# load nimbleEcology in case we forgot previously
library(nimbleEcology)
# data
y <- dipper %>%
  select(year_1981:year_1987) %>%
  as.matrix()
y <- y[ first != ncol(y), ] # get rid of individuals for which
                            # first detection = last capture
```

```r
# NIMBLE code
hmm.phip.nimbleecology <- nimbleCode({
  phi ~ dunif(0, 1) # survival prior
  p ~ dunif(0, 1)    # detection prior
  # likelihood
  for (i in 1:N){
    y[i, first[i]:T] ~ dCJS_ss(probSurvive = phi,
                               probCapture = p,
                               len = T - first[i] + 1)
  }
})
# occasions of first capture
first <- apply(y, 1, function(x) min(which(x !=0)))
# constants
my.constants <- list(N = nrow(y),
                     T = ncol(y),
                     first = first)
# data
my.data <- list(y = y) # 0 = non-detected, 1 = detected
# initial values: marginalized likelihood, hence no latent states
# therefore no need for initial values for latent states
initial.values <- function() list(phi = runif(1,0,1),
                                   p = runif(1,0,1))
# parameters to monitor
parameters.to.save <- c("phi", "p")
# MCMC details
n.iter <- 2500
n.burnin <- 1000
n.chains <- 2
# run NIMBLE
mcmc.phip.nimbleecology <- nimbleMCMC(code = hmm.phip.nimbleecology,
                      constants = my.constants,
                      data = my.data,
                      inits = initial.values,
                      monitors = parameters.to.save,
                      niter = n.iter,
                      nburnin = n.burnin,
                      nchains = n.chains)
```

```
## |-------------|--------------|--------------|--------------|
## |-----------------------------------------------------------|
## |-------------|--------------|--------------|--------------|
## |-----------------------------------------------------------|
# numerical summaries
MCMCsummary(mcmc.phip.nimbleecology, round = 2)
##       mean   sd 2.5%  50% 97.5% Rhat n.eff
## p     0.90 0.03 0.83 0.90  0.94 1.00   668
## phi   0.56 0.02 0.51 0.56  0.61 1.01   723
```

We are left with four models, each saying a different story about the data, with temporal variation or not in either survival or detection probability. How to quantify the most plausible of these four ecological hypotheses? Rendez-vous in the next section.

4.6   Model comparison with WAIC

Which of the four models above is best supported by the data? To address this question, we need to remember that we used all the observed data to fit these models. What really matters is how well each model will perform when predicting future data, a property known as predictive accuracy. A natural measure of predictive accuracy is the likelihood, often referred in the context of model comparison as the predictive density. In practice, however, we do not know the true process or the future data, so the predictive density can only be estimated, and with some bias.

You may have heard about the Akaike Information Criterion (AIC) in the frequentist framework, and the Deviance Information Criterion (DIC) in the Bayesian framework. Here, we will also consider the Widely Applicable Information Criterion or Watanabe Information Criterion (WAIC). AIC, DIC and WAIC each aim to provide an approximation of predictive accuracy.

While AIC is the predictive measure of choice in the frequentist framework for ecologists, DIC has been around for some time for Bayesian applications due to its availability in popular BUGS pieces of software. However, AIC and BIC use a point estimate of the unknown parameters, while we have access to their entire (posterior) distribution with the Bayesian approach. Also, DIC has been shown to misbehave when the posterior distribution is not well summarized by its mean. For a more fully Bayesian approach we would like to use the entire posterior distribution to evaluate the predictive performance, which is exactly what WAIC does.

Conveniently, NIMBLE calculates WAIC for you. The only modification you need to make is to add `WAIC = TRUE` in the call to the `nimbleMCMC()` function. For example, for the CJS model, we write:

```r
mcmc.phitpt <- nimbleMCMC(code = hmm.phitpt,
                          constants = my.constants,
                          data = my.data,
                          inits = initial.values,
                          monitors = parameters.to.save,
                          niter = n.iter,
                          nburnin = n.burnin,
                          nchains = n.chains,
                          WAIC = TRUE)
```

I re-ran the four models to calculate the WAIC value for each of them

```r
data.frame(model = c("both survival & detection constant",
                     "time-dependent survival, constant detection",
                     "constant survival, time-dependent detection",
                     "both survival & detection time-dependent"),
           WAIC = c(mcmc.phip$WAIC$WAIC,
                    mcmc.phitp$WAIC$WAIC,
                    mcmc.phipt$WAIC$WAIC,
                    mcmc.phitpt$WAIC$WAIC))
```

```
##                                               model  WAIC
## 1                both survival & detection constant 266.7
## 2 time-dependent survival, constant detection 273.0
```

```
## 3 constant survival, time-dependent detection 270.9
## 4    both survival & detection time-dependent 308.9
```

Lower values of WAIC imply higher predictive accuracy, therefore we would favor model with constant parameters.

4.7 Goodness of fit

In the previous section, Section 4.6, we compared models between each other based on their predictive accuracy – we assessed their *relative* fit. However, even though we were able to rank these models according to predictive accuracy, it could happen that all models actually had poor predictive performance – this has to do with *absolute* fit.

How to assess the goodness of fit of the CJS model to capture-recapture data?

4.7.1 Posterior predictive checks

In the Bayesian framework, we use posterior predictive checks to assess absolute model fit. The basic idea is to compare the observed data with replicated data generated from the model. If the model is a good fit to the data, then the replicated data predicted from the model should look similar to the observed data. To make the comparison tractable, some summary statistics are generally used. For the CJS model, this is often done with the m-array, which records the number of individuals released in a given year and first recaptured in subsequent years. For examples, see Kéry and Schaub [2011], Paganin and de Valpine [2025], and the NIMBLE website https://r-nimble.org/examples/posterior_pr edictive.html.

Posterior predictive checks are powerful tools. However, there are long-established procedures for evaluating absolute fit and diagnosing violations of specific CJS model assumptions, and it would be a shame to ignore them.

4.7.2 Classical tests

We focus here on two assumptions that have an ecological interpretation: transience and trap-dependence. The transience test evaluates whether newly encountered individuals are as likely to be detected again as previously encountered ones. A signal of transience is often seen as an excess of individuals never seen again. The trap-dependence test evaluates whether individuals that were not detected at a given occasion are as likely to be detected at the next occasion as those that were. Trap-dependence is often seen as an effect of trapping on detection. Although these tests are commonly referred to as the test of transience and the test of trap-dependence, you should keep in mind that transience and trap-dependence are only two possible reasons why these tests may reveal a lack of fit.

These tests are implemented in the package R2ucare, and we illustrate its use with the dipper data.

We load the package R2ucare:

```r
# To install the R2ucare package:
# if(!require(devtools)) install.packages("devtools")
# devtools::install_github("oliviergimenez/R2ucare")
library(R2ucare)
```

We get the capture-recapture data:

```r
# capture-recapture data
dipper <- read_csv("dat/dipper.csv")
# get encounter histories
y <- dipper %>%
  select(year_1981:year_1987) %>%
  as.matrix()
# number of birds with a particular capture-recapture history
size <- rep(1, nrow(y))
```

We may perform a test specifically to assess a transient effect:

```
test3sr(y, size)
## $test3sr
##       stat        df      p_val sign_test
##      1.773     5.000      0.880    -0.054
##
## $details
##    component  stat p_val signed_test   test_perf
## 1          2  0.08 0.778       0.283  Chi-square
## 2          3 0.232  0.63       0.482  Chi-square
## 3          4 0.847 0.358       -0.92  Chi-square
## 4          5 0.288 0.592      -0.537  Chi-square
## 5          6 0.326 0.568       0.571  Chi-square
```

Or trap-dependence:

```
test2ct(y, size)
## $test2ct
##       stat        df      p_val sign_test
##      9.463     4.000      0.051    -1.538
##
## $details
##    component dof  stat p_val signed_test test_perf
## 1          2   1     0     1           0    Fisher
## 2          3   1     0     1           0    Fisher
## 3          4   1     0     1           0    Fisher
## 4          5   1 9.463 0.002      -3.076    Fisher
```

None of these tests is significant, but what to do if these tests are significant? Transience can occur for different reasons. Some animals may simply pass through without belonging to the study population (true transients). Others may reproduce once and then die or permanently disperse, reflecting the cost of first reproduction. In some cases, marking itself may cause individuals to die or leave the area. Whatever the cause, these transient individuals are never detected again after their initial capture, so their probability of local survival is zero after initial capture.

This suggests a way to account for transience by modeling a two-age class effect on survival, where age here refers not to biological age but to time since first capture. The first age class, Age 1, corresponds to newly marked in the interval immediately following capture. This is where transients live: if they are true transients, they will never be seen again because their local survival after marking is effectively 0. The second age class, Age 2+ corresponds to established residents and all subsequent intervals for individuals that remained in the study population beyond the first interval.

We allow survival to differ between these two ages, defining ϕ_1 as the apparent survival in the first interval after initial capture, and ϕ_{2+} as the apparent survival in all subsequent intervals. If there are transients, ϕ_1 will be pulled lower than ϕ_{2+}, since the first interval mixes true residents (who survive locally with probability ϕ_{2+}) and transients (who never survive locally to the next occasion). A useful quantity to compute here is the transient proportion τ, defined as $\tau = 1 - \dfrac{\phi_1}{\phi_{2+}}$.

We will later see in Section 4.8.5 how to encode an age effect on survival, and I will use this approach I just described to account for transience in Section 7.2. Note also that transience can be represented through states in a HMM [see Genovart and Pradel, 2019, for more details].

Trap-dependence may arise in several ways. It can occur when animals are directly affected by trapping, either positively or negatively (true trap-dependence). It may also result from observers visiting some parts of the study area more often after individuals have been detected (preferential sampling), or from habitat heterogeneity, where some patches are more accessible and individuals located there have higher detection probabilities (access bias). Trap-dependence can also reflect individual differences in age, sex, or social status that influence movements or activity patterns and thus detection probability (individual heterogeneity). Finally, it may arise when temporary emigration is non-random, for instance if individuals skip reproduction (skipped reproduction).

In short, trap-dependence designates some form of correlation between detection events. To address this issue, this correlation needs to be accounted for, and we show how to do so in a case study at Section 7.2.

4.7.3   Design considerations

So far, we have considered assumptions related to the model. There are also assumptions tied to the study design. In particular, when we talk about survival, it is always in reference to the study area, and we need to be clear about what this means. What we estimate is usually called *apparent* survival, not *true* survival. Apparent survival is the product of true survival and site fidelity and is therefore always lower than true survival unless fidelity equals one.

Other design assumptions are worth noting, which I only list here. Marks should not be lost, individual identity should be recorded without error (no false positives), and captured animals should represent a random sample of the population.

4.8   Covariates

In the models we have considered so far, survival and detection probabilities may vary over time, but we do not include ecological drivers that might explain these variation. Luckily, in the spirit of generalized linear models, we can make these parameters dependent on external covariates over time, such as environmental conditions for survival or sampling effort for detection.

Besides variation over time, we will also cover individual variation in these parameters, in which for example survival vary according to sex or some phenotypic characteristics (e.g. size or body mass).

Let's illustrate the use of covariates with the dipper example.

4.8.1 Temporal covariates

4.8.1.1 Discrete

A major flood occurred during the 1983 breeding season (from October 1982 to May 1983). Because captures during the breeding season occurred before and after the flood, survival in the two years 1982–1983 and 1983–1984 were likely to be affected. Indeed survival of a species living along and feeding in the river in these two flood years was most likely lower than in nonflood years. Here we will use a discrete or categorical covariate, or a group.

Let's use a covariate `flood` that contains 1s and 2s, indicating whether we are in a flood or nonflood year for each year: 1 if nonflood year, and 2 if flood year.

```r
flood <- c(1, # 1981-1982 (nonflood)
           2, # 1982-1983 (flood)
           2, # 1983-1984 (flood)
           1, # 1984-1985 (nonflood)
           1, # 1985-1986 (nonflood)
           1) # 1986-1987 (nonflood)
```

Then we write the model code:

```r
hmm.phifloodp <- nimbleCode({
  delta[1] <- 1                            # Pr(alive t = 1) = 1
  delta[2] <- 0                            # Pr(dead t = 1) = 0
  for (t in 1:(T-1)){
    gamma[1,1,t] <- phi[flood[t]]          # Pr(alive t -> alive t+1)
    gamma[1,2,t] <- 1 - phi[flood[t]]      # Pr(alive t -> dead t+1)
    gamma[2,1,t] <- 0                      # Pr(dead t -> alive t+1)
    gamma[2,2,t] <- 1                      # Pr(dead t -> dead t+1)
  }
  p ~ dunif(0, 1)          # prior detection
  omega[1,1] <- 1 - p      # Pr(alive t -> non-detected t)
  omega[1,2] <- p          # Pr(alive t -> detected t)
  omega[2,1] <- 1          # Pr(dead t -> non-detected t)
```

```r
  omega[2,2] <- 0           # Pr(dead t -> detected t)
  phi[1] ~ dunif(0, 1)      # prior for survival in nonflood years
  phi[2] ~ dunif(0, 1)      # prior for survival in flood years
  # likelihood
  for (i in 1:N){
    z[i,first[i]] ~ dcat(delta[1:2])
    for (j in (first[i]+1):T){
      z[i,j] ~ dcat(gamma[z[i,j-1], 1:2, j-1])
      y[i,j] ~ dcat(omega[z[i,j], 1:2])
    }
  }
})
```

We use nested indexing above when specifying survival in the transition matrix. E.g. for year $t = 2$, `phi[flood[t]]` gives `phi[flood[2]]` which will be `phi[2]` because `flood[2]` is 2 that (a flood year).

Let's provide the constants in a list:

```r
my.constants <- list(N = nrow(y),
                     T = ncol(y),
                     first = first,
                     flood = flood)
```

Now a function to generate initial values:

```r
initial.values <- function() list(phi = runif(2,0,1),
                                   p = runif(1,0,1),
                                   z = zinits)
```

The parameters to be monitored:

```r
parameters.to.save <- c("p", "phi")
```

The MCMC details:

```
n.iter <- 5000
n.burnin <- 1000
n.chains <- 2
```

We're all set, and we run NIMBLE:

```
mcmc.phifloodp <- nimbleMCMC(code = hmm.phifloodp,
                             constants = my.constants,
                             data = my.data,
                             inits = initial.values,
                             monitors = parameters.to.save,
                             niter = n.iter,
                             nburnin = n.burnin,
                             nchains = n.chains)
```

The numerical summaries are given by:

```
MCMCsummary(mcmc.phifloodp, round = 2)
```

```
##          mean   sd 2.5%  50% 97.5% Rhat n.eff
## p        0.89 0.03 0.83 0.90  0.95    1   609
## phi[1]   0.61 0.03 0.55 0.61  0.67    1  1474
## phi[2]   0.47 0.04 0.38 0.47  0.56    1  1700
```

Having a look to the numerical summaries, we see that as expected, survival in flood years (`phi[2]`) is much lower than survival in non-flood years (`phi[1]`). You could formally test this difference by considering the difference `phi[1] - phi[2]`. Alternatively, this can be done afterwards and calculating the probability that this difference is positive (or `phi[1] > phi[2]`). Using a single chain for convenience, we do:

```
phi1 <- mcmc.phifloodp$chain1[,'phi[1]']
phi2 <- mcmc.phifloodp$chain1[,'phi[2]']
mean(phi1 - phi2 > 0)
## [1] 0.9952
```

An important point here is that we formulated an ecological hypothesis which we translated into a model. The next step would consist in calculating WAIC for this model and compare it with that of the four other model we have fitted so far (see Section 4.6).

Another method to include a discrete covariate consists in considering the effect of the difference between its levels. For example, we could consider survival in nonflood years as the reference and test the difference in survival with flood years.

To do so, we write survival as a linear function of our covariate on some scale, e.g. $\text{logit}(\phi_t) = \beta_1 + \beta_2 \, \text{flood}_t$ where the β's are regression coefficients we need to estimate (intercept β_1 and slope β_2), and $\text{logit}(x) = \log\left(\dfrac{x}{1-x}\right)$ is the logit function. The logit function lives between $-\infty$ and $+\infty$, and sends values between 0 and 1 onto the real line. For example $\log(0.2/(1-0.2)) = -1.39$, $\log(0.5/(1-0.5)) = 0$ and $\log(0.8/(1-0.8)) = 1.39$. Why use the logit function? Because survival is a probability bounded between 0 and 1, and if we used directly $\phi_t = \beta_1 + \beta_2 \, \text{flood}_t$ we might end up with estimates for the regression coefficients that make survival out of bound. Therefore, we consider survival as a linear function of our covariates on the scale of the logit function, working on the real line, then back-transform using the inverse-logit (reciprocal function) to obtain survival on its natural scale. The inverse-logit function is $\text{logit}^{-1}(x) = \dfrac{\exp(x)}{1+\exp(x)} = \dfrac{1}{1+\exp(-x)}$. The logit function is often called a link function like in generalized linear models. Other link functions exist, we'll see an example in Section 5.4.2.

Another point of attention is the prior we will assign to regression coefficients. We no longer assign a prior to survival directly like previously, but we need to assign prior to the β's which will induce some prior on survival. And sometimes, your priors on the regression coefficients are non-informative, but the prior on survival is not. Consider for example the case of a single intercept with no covariate. If you assign as a prior to this regression coefficient a normal distribution with mean 0 and large standard deviation (left figure below), which would be my first reflex, then you end up with a very informative prior on survival with a bathtub shape, putting much importance on low and high values (right figure below):

```r
# 1000 random values from a N(0,10)
intercept <- rnorm(1000, mean = 0, sd = 10)
# plogis() is the inverse-logit function in R
survival <- plogis(intercept)
df <- data.frame(intercept = intercept, survival = survival)
plot1 <- df %>%
  ggplot(aes(x = intercept)) +
  geom_density() +
  labs(x = "prior on intercept")
plot2 <- df %>%
  ggplot(aes(x = survival)) +
  geom_density() +
  labs(x = "prior on survival")
plot1 + plot2
```

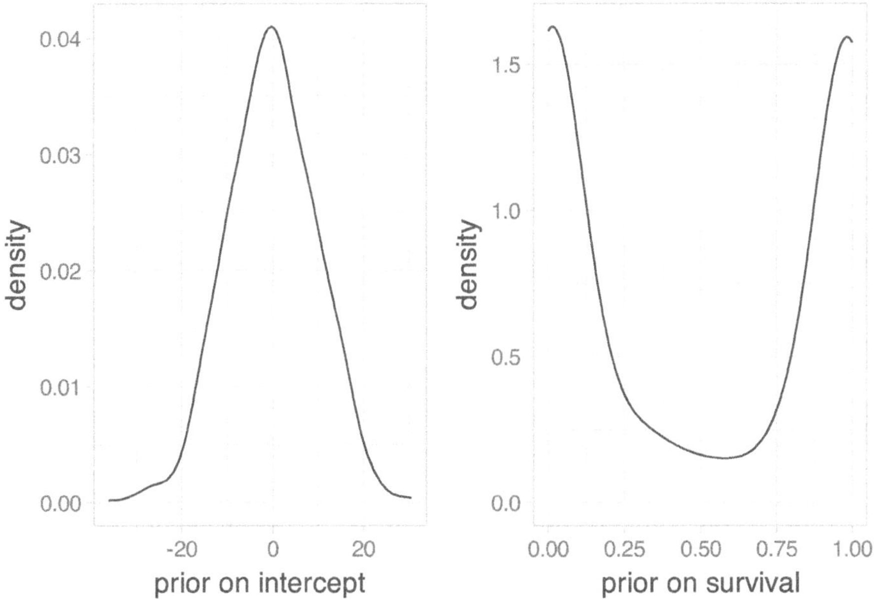

Now if you go for a lower standard deviation for the intercept prior (left figure below), e.g. 1.5, the prior on survival is non-informative, looking like a uniform distribution between 0 and 1 (right figure below):

```r
set.seed(123)
# 1000 random values from a N(0,1.5)
intercept <- rnorm(1000, mean = 0, sd = 1.5)
# plogis() is the inverse-logit function in R
survival <- plogis(intercept)
df <- data.frame(intercept = intercept, survival = survival)
plot1 <- df %>%
  ggplot(aes(x = intercept)) +
  geom_density() +
  labs(x = "prior on intercept")
plot2 <- df %>%
  ggplot(aes(x = survival)) +
  geom_density() +
  labs(x = "prior on survival")
plot1 + plot2
```

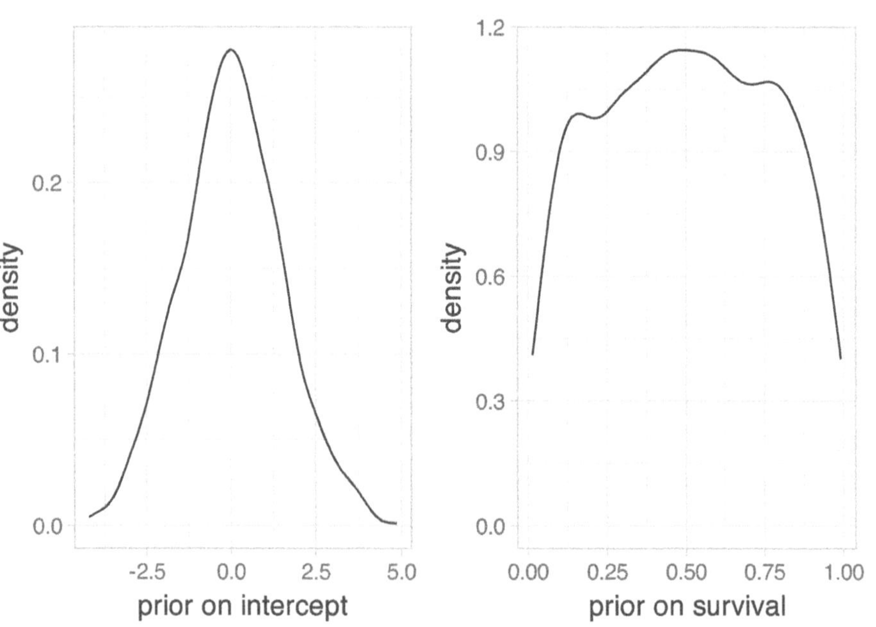

Now let's go back to our model. We first define our flood covariate with 0 if nonflood year, and 1 if flood year:

```r
flood <- c(0, # 1981-1982 (nonflood)
           1, # 1982-1983 (flood)
           1, # 1983-1984 (flood)
           0, # 1984-1985 (nonflood)
           0, # 1985-1986 (nonflood)
           0) # 1986-1987 (nonflood)
```

Then we write the NIMBLE code:

```r
hmm.phifloodp <- nimbleCode({
  delta[1] <- 1                    # Pr(alive t = 1) = 1
  delta[2] <- 0                    # Pr(dead t = 1) = 0
  for (t in 1:(T-1)){
    logit(phi[t]) <- beta[1] + beta[2] * flood[t]
    gamma[1,1,t] <- phi[t]         # Pr(alive t -> alive t+1)
    gamma[1,2,t] <- 1 - phi[t]     # Pr(alive t -> dead t+1)
    gamma[2,1,t] <- 0              # Pr(dead t -> alive t+1)
    gamma[2,2,t] <- 1              # Pr(dead t -> dead t+1)
  }
  p ~ dunif(0, 1)                  # prior detection
  omega[1,1] <- 1 - p              # Pr(alive t -> non-detected t)
  omega[1,2] <- p                  # Pr(alive t -> detected t)
  omega[2,1] <- 1                  # Pr(dead t -> non-detected t)
  omega[2,2] <- 0                  # Pr(dead t -> detected t)
  beta[1] ~ dnorm(0, sd = 1.5)     # prior intercept
  beta[2] ~ dnorm(0, sd = 1.5)     # prior slope
  # likelihood
  for (i in 1:N){
    z[i,first[i]] ~ dcat(delta[1:2])
    for (j in (first[i]+1):T){
      z[i,j] ~ dcat(gamma[z[i,j-1], 1:2, j-1])
      y[i,j] ~ dcat(omega[z[i,j], 1:2])
    }
  }
})
```

We wrote $\text{logit}(\phi_t) = \beta_1 + \beta_2\,\text{flood}_t$, meaning that survival in nonflood years ($\text{flood}_t = 0$) is $\text{logit}(\phi_t) = \beta_1$ and survival in flood years ($\text{flood}_t = 1$) is $\text{logit}(\phi_t) = \beta_1 + \beta_2$. We see that β_1 is survival in nonflood years (on the logit scale) and β_2 is the difference between survival in flood years and survival in nonflood years (again, on the logit scale). In passing we assigned the same prior for both β_1 and β_2 but in certain situations, we might think twice before doing that as β_2 is a difference between two survival probabilities (on the logit scale).

Let's put our constants in a list:

```r
my.constants <- list(N = nrow(y),
                     T = ncol(y),
                     first = first,
                     flood = flood)
```

Then our function for generating initial values:

```r
initial.values <- function() list(beta = rnorm(2,0,1),
                                   p = runif(1,0,1),
                                   z = zinits)
```

And the parameters to be monitored:

```r
parameters.to.save <- c("beta", "p", "phi")
```

Finaly, we run NIMBLE:

```r
mcmc.phifloodp <- nimbleMCMC(code = hmm.phifloodp,
                             constants = my.constants,
                             data = my.data,
                             inits = initial.values,
                             monitors = parameters.to.save,
                             niter = n.iter,
                             nburnin = n.burnin,
                             nchains = n.chains)
```

You may check that we get the same numerical summaries as above for survival in nonflood years (phi[1], phi[4], phi[5] and phi[6]) and flood years (phi[2] and phi[3]):

```
MCMCsummary(mcmc.phifloodp, round = 2)
```

```
##             mean   sd   2.5%   50%  97.5%  Rhat n.eff
## beta[1]   0.44  0.13   0.17  0.44   0.69  1.01   725
## beta[2] -0.54  0.22 -0.96 -0.54 -0.11  1.00   770
## p          0.89  0.03   0.83  0.89   0.94  1.00   631
## phi[1]    0.61  0.03   0.54  0.61   0.67  1.01   724
## phi[2]    0.47  0.04   0.39  0.47   0.56  1.00  1597
## phi[3]    0.47  0.04   0.39  0.47   0.56  1.00  1597
## phi[4]    0.61  0.03   0.54  0.61   0.67  1.01   724
## phi[5]    0.61  0.03   0.54  0.61   0.67  1.01   724
## phi[6]    0.61  0.03   0.54  0.61   0.67  1.01   724
```

You may also check how to go from the β's to the survival probabilities ϕ. Let's get the draws from the posterior distribution of the β's first:

```
beta1 <- c(mcmc.phifloodp$chain1[,'beta[1]'], # beta1 chain 1
           mcmc.phifloodp$chain2[,'beta[1]']) # beta1 chain 2
beta2 <- c(mcmc.phifloodp$chain1[,'beta[2]'], # beta2 chain 1
           mcmc.phifloodp$chain2[,'beta[2]']) # beta2 chain 2
```

Then apply the inverse-logit function to get survival in nonflood years, e.g. its posterior mean and credible interval:

```
mean(plogis(beta1))
## [1] 0.6066
quantile(plogis(beta1), probs = c(2.5, 97.5)/100)
##    2.5%   97.5%
## 0.5435 0.6651
```

Same thing for survival in flood years:

```
mean(plogis(beta1 + beta2))
## [1] 0.4744
quantile(plogis(beta1 + beta2), probs = c(2.5, 97.5)/100)
##    2.5%   97.5%
## 0.3895 0.5606
```

4.8.1.2 Continuous

Instead of a discrete covariate varying over time, we may want to consider a continuous covariate, say x_t, through logit$(\phi_t) = \beta_1 + \beta_2 x_t$. For example, let's investigate the effect of water flow on dipper survival, which should reflect the flood that occurred during the 1983 breeding season.

We build a covariate with water flow in liters per second measured during the March to May period each year, starting with year 1982:

```
# water flow in L/s
water_flow <- c(443,   # March-May 1982
                1114,  # March-May 1983
                529,   # March-May 1984
                434,   # March-May 1985
                627,   # March-May 1986
                466)   # March-May 1987
```

We do not need water flow in 1981 because we will write the probability ϕ_t of being alive in year $t + 1$ given a bird was alive in year t as a linear function of the water flow in year $t + 1$.

You may have noticed the high value of water flow for 1983, twice as much as in the other years, corresponding to the flood. Importantly, we standardize our covariate to improve convergence:

```r
water_flow_st <- (water_flow - mean(water_flow))/sd(water_flow)
```

Now we write the model code:

```r
hmm.phiflowp <- nimbleCode({
  delta[1] <- 1                        # Pr(alive t = 1) = 1
  delta[2] <- 0                        # Pr(dead t = 1) = 0
  for (t in 1:(T-1)){
    logit(phi[t]) <- beta[1] + beta[2] * flow[t]
    gamma[1,1,t] <- phi[t]          # Pr(alive t -> alive t+1)
    gamma[1,2,t] <- 1 - phi[t]      # Pr(alive t -> dead t+1)
    gamma[2,1,t] <- 0              # Pr(dead t -> alive t+1)
    gamma[2,2,t] <- 1              # Pr(dead t -> dead t+1)
  }
  p ~ dunif(0, 1)                      # prior detection
  omega[1,1] <- 1 - p                  # Pr(alive t -> non-detected t)
  omega[1,2] <- p                      # Pr(alive t -> detected t)
  omega[2,1] <- 1                      # Pr(dead t -> non-detected t)
  omega[2,2] <- 0                      # Pr(dead t -> detected t)
  beta[1] ~ dnorm(0, 1.5)              # prior intercept
  beta[2] ~ dnorm(0, 1.5)              # prior slope
  # likelihood
  for (i in 1:N){
    z[i,first[i]] ~ dcat(delta[1:2])
    for (j in (first[i]+1):T){
      z[i,j] ~ dcat(gamma[z[i,j-1], 1:2, j-1])
      y[i,j] ~ dcat(omega[z[i,j], 1:2])
    }
  }
})
```

We put the constants in a list:

```r
my.constants <- list(N = nrow(y),
                     T = ncol(y),
                     first = first,
                     flow = water_flow_st)
```

Initial values as usual:

```r
initial.values <- function() list(beta = rnorm(2,0,1),
                                   p = runif(1,0,1),
                                   z = zinits)
```

And parameters to be monitored:

```r
parameters.to.save <- c("beta", "p", "phi")
```

Eventually, we run NIMBLE:

```r
mcmc.phiflowp <- nimbleMCMC(code = hmm.phiflowp,
                            constants = my.constants,
                            data = my.data,
                            inits = initial.values,
                            monitors = parameters.to.save,
                            niter = n.iter,
                            nburnin = n.burnin,
                            nchains = n.chains)
```

We can have a look to the results through a caterpillar plot of the regression parameters:

```r
MCMCplot(object = mcmc.phiflowp, params = "beta")
```

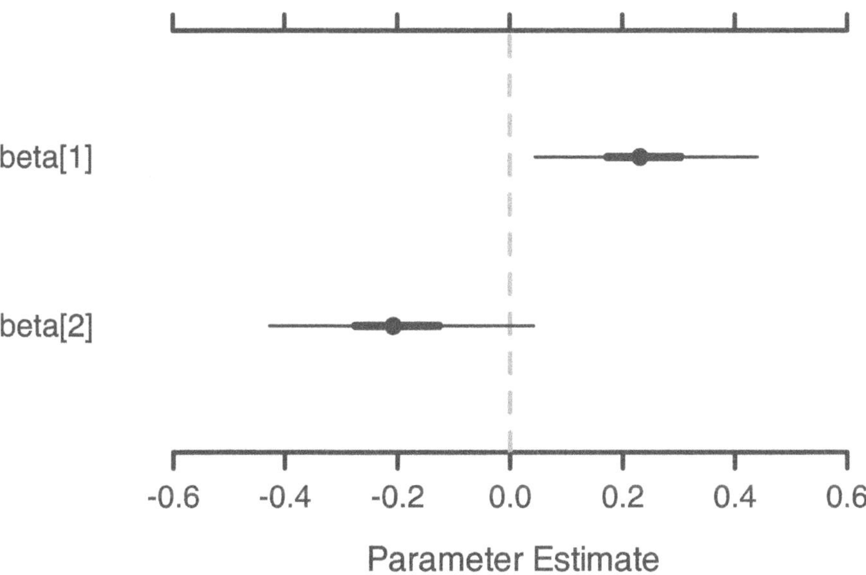

The posterior distribution of the slope (`beta[2]`) is centered on negative values, suggesting that as water flow increases, survival decreases.

Let's inspect the time-dependent survival probability:

```r
MCMCplot(object = mcmc.phiflowp, params = "phi", ISB = TRUE)
```

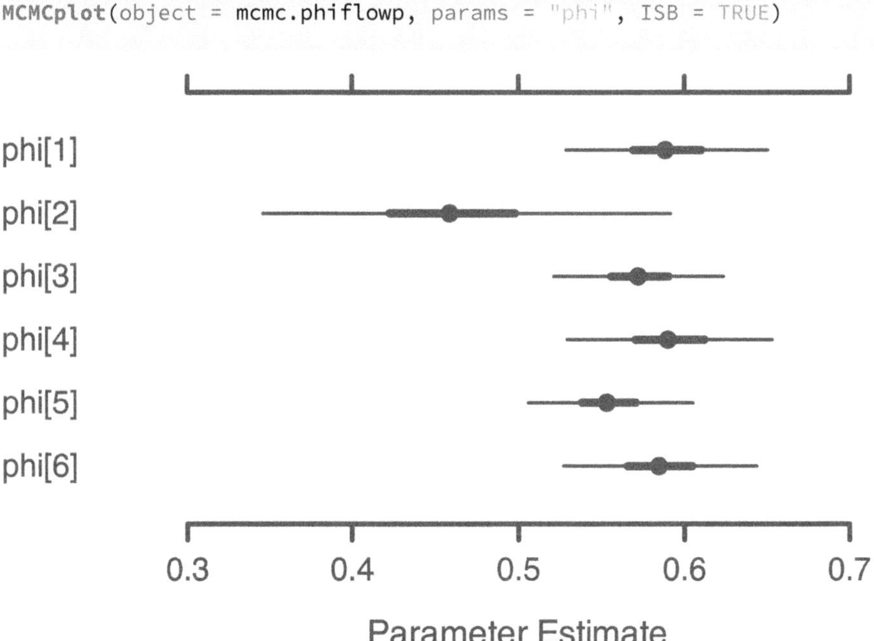

Survival between 1982 and 1983 (phi[2]) was greatly affected and much lower than on average. This decrease corresponds to the high water flow in 1983 and the flood. These results are in line with our previous findings obtained by considering a discrete covariate for nonflood vs. flood years.

4.8.2 Individual covariates

In the previous section, we learnt how to explain temporal heterogeneity in survival and detection. Heterogeneity could also originate from individual differences between animals. You may think of a difference in survival between males and females for a discrete covariate example, or size and body mass for examples of a continuous covariate. Let's illustrate both discrete and continuous covariates on the dipper.

4.8.2.1 Discrete

We first consider a covariate sex that contains 1's and 2's indicating the sex of each bird: 1 if male, and 2 if female. We implement the model with sex effect using nested indexing, similarly to the model with flood vs. nonflood years. The section of the NIMBLE code that needs to be amended is:

```
...
for (i in 1:N){
  gamma[1,1,i] <- phi[sex[i]]     # Pr(alive t -> alive t+1)
  gamma[1,2,i] <- 1 - phi[sex[i]] # Pr(alive t -> dead t+1)
  gamma[2,1,i] <- 0               # Pr(dead t -> alive t+1)
  gamma[2,2,i] <- 1               # Pr(dead t -> dead t+1)
}
phi[1] ~ dunif(0,1) # male survival
phi[2] ~ dunif(0,1) # female survival
...
```

After running NIMBLE, we get:

```
MCMCsummary(object = mcmc.phisexp.ni, round = 2)
```

```
##      mean   sd 2.5%  50% 97.5% Rhat n.eff
## p    0.90 0.03 0.84 0.90  0.95    1   643
```

```
## phi[1] 0.57 0.03 0.50 0.57  0.64    1  1668
## phi[2] 0.55 0.03 0.48 0.55  0.62    1  1482
```

Male survival (`phi[1]`) looks very similar to female survival (`phi[2]`).

4.8.2.2 Continuous

Besides discrete individual covariates, you might want to have continuous individual covariates, e.g. wing length in the dipper example. Note that we're considering an individual trait that takes the same value whatever the occasion. We consider wing length here, and more precisely its measurement at first detection. We first standardize the covariate:

```r
wing.length.st <- as.vector(scale(dipper$wing_length))
head(wing.length.st)
## [1]  0.7581 -0.8671  0.5260 -1.5637 -1.3315  1.2225
```

Now we write the model:

```r
hmm.phiwlp <- nimbleCode({
    p ~ dunif(0, 1)                  # prior detection
    omega[1,1] <- 1 - p             # Pr(alive t -> non-detected t)
    omega[1,2] <- p                 # Pr(alive t -> detected t)
    omega[2,1] <- 1                 # Pr(dead t -> non-detected t)
    omega[2,2] <- 0                 # Pr(dead t -> detected t)
  for (i in 1:N){
    logit(phi[i]) <- beta[1] + beta[2] * winglength[i]
    gamma[1,1,i] <- phi[i]          # Pr(alive t -> alive t+1)
    gamma[1,2,i] <- 1 - phi[i]      # Pr(alive t -> dead t+1)
    gamma[2,1,i] <- 0               # Pr(dead t -> alive t+1)
    gamma[2,2,i] <- 1               # Pr(dead t -> dead t+1)
  }
  beta[1] ~ dnorm(mean = 0, sd = 1.5)
  beta[2] ~ dnorm(mean = 0, sd = 1.5)
  delta[1] <- 1                     # Pr(alive t = 1) = 1
  delta[2] <- 0                     # Pr(dead t = 1) = 0
  # likelihood
```

```
  for (i in 1:N){
    z[i,first[i]] ~ dcat(delta[1:2])
    for (j in (first[i]+1):T){
      z[i,j] ~ dcat(gamma[z[i,j-1], 1:2, i])
      y[i,j] ~ dcat(omega[z[i,j], 1:2])
    }
  }
})
```

We put the constants in a list:

```
my.constants <- list(N = nrow(y),
                     T = ncol(y),
                     first = first,
                     winglength = wing.length.st)
```

We write a function for generating initial values:

```
initial.values <- function() list(beta = rnorm(2,0,1),
                                   p = runif(1,0,1),
                                   z = zinits)
```

And we run NIMBLE:

```
mcmc.phiwlp <- nimbleMCMC(code = hmm.phiwlp,
                          constants = my.constants,
                          data = my.data,
                          inits = initial.values,
                          monitors = parameters.to.save,
                          niter = n.iter,
                          nburnin = n.burnin,
                          nchains = n.chains)
```

Let's inspect the numerical summaries for the regression parameters:

```
MCMCsummary(mcmc.phiwlp, params = "beta", round = 2)
```

```
##            mean   sd  2.5%    50% 97.5% Rhat n.eff
## beta[1]   0.25 0.10  0.04   0.25  0.45    1  1472
## beta[2]  -0.02 0.09 -0.20  -0.02  0.17    1  1555
```

Wing length does not seem to explain much individual-to-individual variation in survival – the posterior distribution of the slope (`beta[2]`) is centered on 0 as we can see from the credible interval.

Let's plot the relationship between survival and wing length. First, we gather the values generated from the posterior distribution of the regression parameters in the two chains:

```
beta1 <- c(mcmc.phiwlp$chain1[,'beta[1]'], # intercept, chain 1
           mcmc.phiwlp$chain2[,'beta[1]']) # intercept, chain 2
beta2 <- c(mcmc.phiwlp$chain1[,'beta[2]'], # slope, chain 1
           mcmc.phiwlp$chain2[,'beta[2]']) # slope, chain 2
```

Then we define a grid of values for wing length, and predict survival for each MCMC iteration:

```
predicted_survival <- matrix(NA,
                             nrow = length(beta1),
                             ncol = length(my.constants$winglength))
for (i in 1:length(beta1)){
  for (j in 1:length(my.constants$winglength)){
    predicted_survival[i,j] <- plogis(beta1[i] +
                        beta2[i] * my.constants$winglength[j])
  }
}
```

Now we calculate posterior mean and the credible interval (note the ordering):

```
mean_survival <- apply(predicted_survival, 2, mean)
lci <- apply(predicted_survival, 2, quantile, prob = 2.5/100)
```

```r
uci <- apply(predicted_survival, 2, quantile, prob = 97.5/100)
ord <- order(my.constants$winglength)
df <- data.frame(wing_length = my.constants$winglength[ord],
                 survival = mean_survival[ord],
                 lci = lci[ord],
                 uci = uci[ord])
```

Now time to visualize:

```r
df %>%
  ggplot() +
  aes(x = wing_length, y = survival) +
  geom_line() +
  geom_ribbon(aes(ymin = lci, ymax = uci),
              fill = "grey70",
              alpha = 0.5) +
  ylim(0,1) +
  labs(x = "wing length", y = "estimated survival")
```

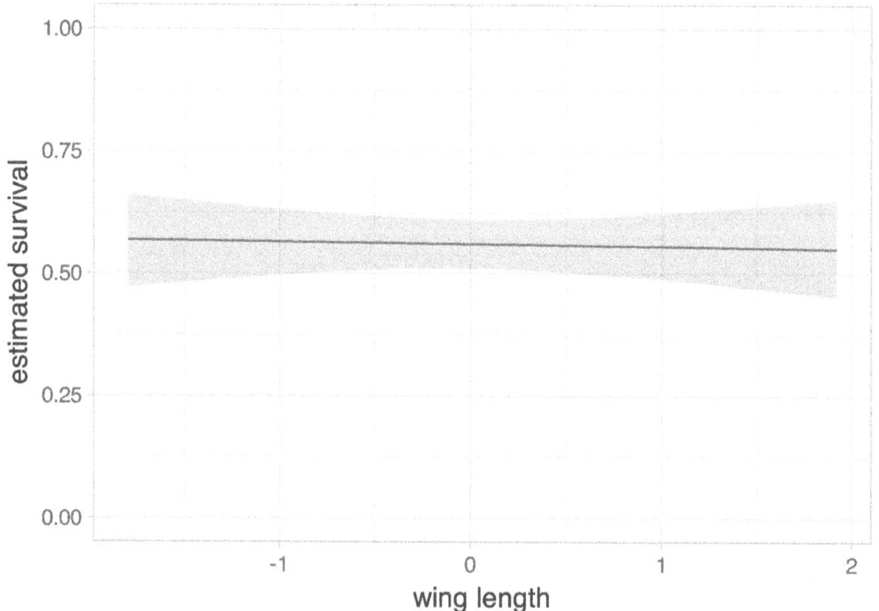

The flat relationship between survival and wing length is confirmed.

4.8.3   Several covariates

You may wish to have an effect of both sex and wing length in a model.
Let's consider an additive effect of both covariates. We use a covariate
sex that takes value 0 if male, and 1 if female. We have $\text{logit}(\phi_i) =$
$\beta_1 + \beta_2\text{sex}_i + \beta_3\text{winglength}_i$ for bird i, so that male survival is $\beta_1 +$
$\beta_3\text{winglength}_i$ and female survival is $\beta_1 + \beta_2 + \beta_3\text{winglength}_i$ (both
on the logit scale). The relationship between survival and wing length
is parallel between males and females, on the logit scale, and the gap
between the two is measured by β_2 (hence the term *additive* effect).

The NIMBLE code is:

```r
hmm.phisexwlp <- nimbleCode({
  p ~ dunif(0, 1)                    # prior detection
  omega[1,1] <- 1 - p                # Pr(alive t -> non-detected t)
  omega[1,2] <- p                    # Pr(alive t -> detected t)
  omega[2,1] <- 1                    # Pr(dead t -> non-detected t)
  omega[2,2] <- 0                    # Pr(dead t -> detected t)
  for (i in 1:N){
    logit(phi[i])<-beta[1]+beta[2]*sex[i]+beta[3]*winglength[i]
    gamma[1,1,i] <- phi[i]           # Pr(alive t -> alive t+1)
    gamma[1,2,i] <- 1 - phi[i]       # Pr(alive t -> dead t+1)
    gamma[2,1,i] <- 0                # Pr(dead t -> alive t+1)
    gamma[2,2,i] <- 1                # Pr(dead t -> dead t+1)
  }
  beta[1] ~ dnorm(mean = 0, sd = 1.5) # intercept male
  beta[2] ~ dnorm(mean = 0, sd = 1.5) # diff bw male and female
  beta[3] ~ dnorm(mean = 0, sd = 1.5) # slope wing length
  delta[1] <- 1                      # Pr(alive t = 1) = 1
  delta[2] <- 0                      # Pr(dead t = 1) = 0
  # likelihood
  for (i in 1:N){
    z[i,first[i]] ~ dcat(delta[1:2])
    for (j in (first[i]+1):T){
      z[i,j] ~ dcat(gamma[z[i,j-1], 1:2, i])
      y[i,j] ~ dcat(omega[z[i,j], 1:2])
    }
```

```
    }
})
```

We put constants and data in lists:

```
first <- apply(y, 1, function(x) min(which(x !=0)))
wing.length.st <- as.vector(scale(dipper$wing_length))
my.constants <- list(N = nrow(y),
                     T = ncol(y),
                     first = first,
                     winglength = wing.length.st,
                     sex = if_else(dipper$sex == "M", 0, 1))
my.data <- list(y = y + 1)
```

We write a function to generate initial values:

```
zinits <- y
zinits[zinits == 0] <- 1
initial.values <- function() list(beta = rnorm(3,0,2),
                                   p = runif(1,0,1),
                                   z = zinits)
```

We specify the parameters to be monitored:

```
parameters.to.save <- c("beta", "p")
```

The MCMC details (note that we need to increase the number of iterations to achieve satisfying effective sample sizes):

```
n.iter <- 5000*4
n.burnin <- 1000
n.chains <- 2
```

And now we run NIMBLE:

```
mcmc.phisexwlp <- nimbleMCMC(code = hmm.phisexwlp,
                             constants = my.constants,
                             data = my.data,
                             inits = initial.values,
                             monitors = parameters.to.save,
                             niter = n.iter,
                             nburnin = n.burnin,
                             nchains = n.chains)
```

Let's display numerical summaries for all parameters:

```
MCMCsummary(mcmc.phisexwlp, round = 2)
```

```
##            mean    sd  2.5%    50% 97.5% Rhat n.eff
## beta[1]    0.52  0.24  0.05   0.52  0.99    1   466
## beta[2]   -0.53  0.43 -1.37  -0.53  0.30    1   447
## beta[3]   -0.25  0.21 -0.65  -0.25  0.16    1   530
## p          0.90  0.03  0.83   0.90  0.95    1  3141
```

The slope `beta[3]` is the same for both males and females. Although its posterior mean is negative, its credible interval suggests that its posterior distribution largely encompasses 0, therefore a very weak signal, if any.

Let's visualize survival as a function of wing length for both sexes. First we put together the values from the two chains we generated in the posterior distributions of the regression parameters:

```
beta1 <- c(mcmc.phisexwlp$chain1[,'beta[1]'], # beta1 chain 1
           mcmc.phisexwlp$chain2[,'beta[1]']) # beta1 chain 2
beta2 <- c(mcmc.phisexwlp$chain1[,'beta[2]'], # beta2 chain 1
           mcmc.phisexwlp$chain2[,'beta[2]']) # beta2 chain 2
beta3 <- c(mcmc.phisexwlp$chain1[,'beta[3]'], # beta3 chain 1
           mcmc.phisexwlp$chain2[,'beta[3]']) # beta3 chain 2
```

We get survival estimates for each MCMC iteration:

```r
predicted_survivalM <- matrix(NA, nrow = length(beta1),
                              ncol=length(my.constants$winglength))
predicted_survivalF <- matrix(NA, nrow = length(beta1),
                              ncol=length(my.constants$winglength))
for (i in 1:length(beta1)){
  for (j in 1:length(my.constants$winglength)){
    predicted_survivalM[i,j] <- plogis(beta1[i] +
                        beta3[i]*my.constants$winglength[j])
    predicted_survivalF[i,j] <- plogis(beta1[i] +
                        beta2[i] +
                        beta3[i]*my.constants$winglength[j])
  }
}
```

From here, we may calculate posterior mean and credible intervals:

```r
mean_survivalM <- apply(predicted_survivalM, 2, mean)
lciM <- apply(predicted_survivalM, 2, quantile, prob = 2.5/100)
uciM <- apply(predicted_survivalM, 2, quantile, prob = 97.5/100)
mean_survivalF <- apply(predicted_survivalF, 2, mean)
lciF <- apply(predicted_survivalF, 2, quantile, prob = 2.5/100)
uciF <- apply(predicted_survivalF, 2, quantile, prob = 97.5/100)
ord <- order(my.constants$winglength)
df <- data.frame(wing_length = c(my.constants$winglength[ord],
                                 my.constants$winglength[ord]),
                 survival = c(mean_survivalM[ord],
                              mean_survivalF[ord]),
                 lci = c(lciM[ord],lciF[ord]),
                 uci = c(uciM[ord],uciF[ord]),
                 sex = c(rep("male", length(mean_survivalM)),
                         rep("female", length(mean_survivalF))))
```

Now on a plot:

```r
df %>%
  ggplot() +
```

```
aes(x = wing_length, y = survival, color = sex) +
geom_line() +
geom_ribbon(aes(ymin = lci, ymax = uci, fill = sex), alpha = 0.5) +
ylim(0,1) +
labs(x = "wing length",
     y = "estimated survival",
     color = "", fill = "")
```

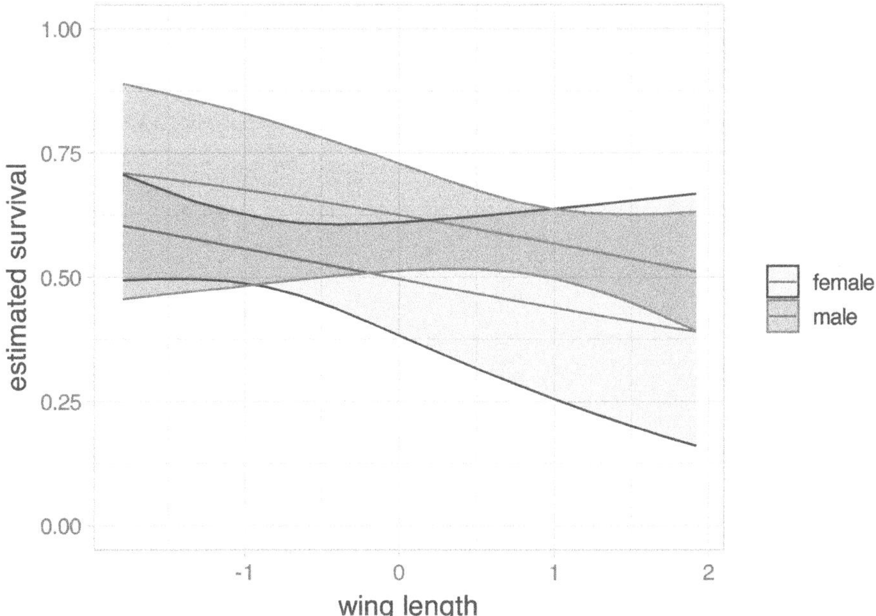

Note that the two curves are not exactly parallel because we back-transformed the linear part of the relationship between survival and wing length. You may check that parallelism occurs on the logit scale:

```
predicted_lsurvivalM <- matrix(NA, nrow = length(beta1),
                               ncol = length(my.constants$winglength))
predicted_lsurvivalF <- matrix(NA, nrow = length(beta1),
                               ncol = length(my.constants$winglength))
for (i in 1:length(beta1)){
  for (j in 1:length(my.constants$winglength)){
    predicted_lsurvivalM[i,j] <- beta1[i] +
      beta3[i] * my.constants$winglength[j]
    predicted_lsurvivalF[i,j] <- beta1[i] + beta2[i] +
```

```r
      beta3[i] * my.constants$winglength[j]
  }
}
mean_lsurvivalM <- apply(predicted_lsurvivalM, 2, mean)
mean_lsurvivalF <- apply(predicted_lsurvivalF, 2, mean)
ord <- order(my.constants$winglength)
df <- data.frame(wing_length = c(my.constants$winglength[ord],
                                 my.constants$winglength[ord]),
             survival = c(mean_lsurvivalM[ord],
                          mean_lsurvivalF[ord]),
             sex = c(rep("male", length(mean_lsurvivalM)),
                     rep("female", length(mean_lsurvivalF))))
df %>%
  ggplot() +
  aes(x = wing_length, y = survival, color = sex) +
  geom_line() +
  ylim(-2,2) +
  labs(x = "wing length",
       y = "estimated survival (on the logit scale)",
       color = "")
```

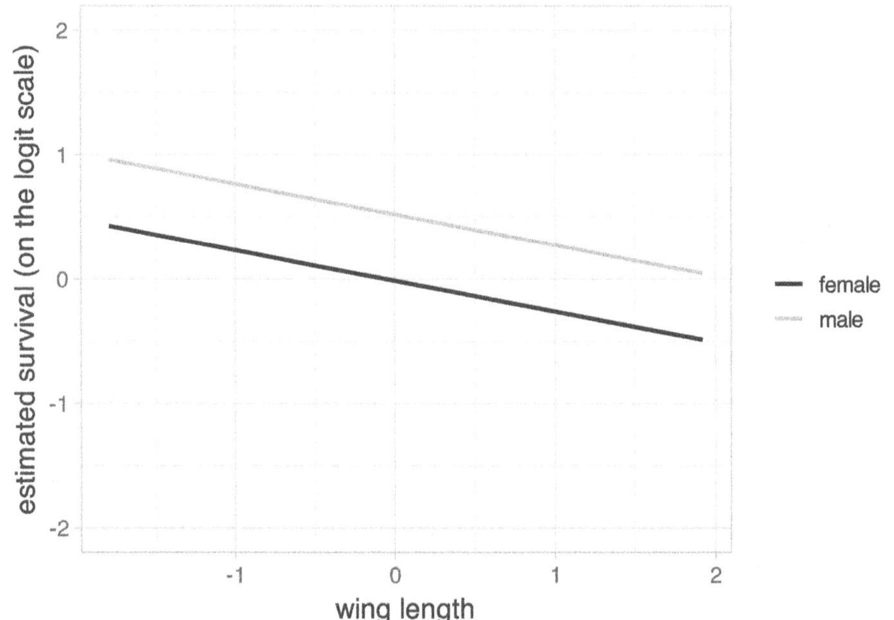

4.8.4 Random effects

If individual variation in survival is not fully explained by our covariates, we may add random effects through $\text{logit}(\phi_i) = \beta_1 + \beta_2 x_i + \varepsilon_i$ where $\varepsilon_i \sim N(0, \sigma^2)$. We consider individual variation in this section, but the reasoning holds for temporal variation. In essence, we are treating the individual survival probabilities ϕ_i as a sample from a population of survival probabilities, and we assume a normal distribution with mean the linear component (with or without a covariate) and standard deviation σ. The variation that is unexplained by the covariate x_i is measured by the variation σ in the residuals ε_i.

Why is it important? Ignoring individual heterogeneity generated by individuals having contrasted performances over life may mask senescence or hamper the understanding of life history trade-offs. Overall, failing to incorporate unexplained residual variance may induce bias in parameter estimates and lead to detecting an effect of the covariate more often than it should be.

Let's go back to our dipper example with wing length as a covariate, and write the NIMBLE code:

```r
hmm.phiwlrep <- nimbleCode({
    p ~ dunif(0, 1)                   # prior detection
    omega[1,1] <- 1 - p              # Pr(alive t -> non-detected t)
    omega[1,2] <- p                 # Pr(alive t -> detected t)
    omega[2,1] <- 1                 # Pr(dead t -> non-detected t)
    omega[2,2] <- 0                 # Pr(dead t -> detected t)
  for (i in 1:N){
    logit(phi[i]) <- beta[1] + beta[2] * winglength[i] + eps[i]
    eps[i] ~ dnorm(mean = 0, sd = sdeps)
    gamma[1,1,i] <- phi[i]        # Pr(alive t -> alive t+1)
    gamma[1,2,i] <- 1 - phi[i]    # Pr(alive t -> dead t+1)
    gamma[2,1,i] <- 0            # Pr(dead t -> alive t+1)
    gamma[2,2,i] <- 1            # Pr(dead t -> dead t+1)
  }
  beta[1] ~ dnorm(mean = 0, sd = 1.5)
  beta[2] ~ dnorm(mean = 0, sd = 1.5)
```

```r
  sdeps ~ dunif(0, 10)
  delta[1] <- 1                        # Pr(alive t = 1) = 1
  delta[2] <- 0                        # Pr(dead t = 1) = 0
  # likelihood
  for (i in 1:N){
    z[i,first[i]] ~ dcat(delta[1:2])
    for (j in (first[i]+1):T){
      z[i,j] ~ dcat(gamma[z[i,j-1], 1:2, i])
      y[i,j] ~ dcat(omega[z[i,j], 1:2])
    }
  }
})
```

The prior on the standard deviation of the random effect is uniform
between 0 and 10.

We now write a function for generating initial values:

```r
initial.values <- function() list(beta = rnorm(2,0,1.5),
                                   sdeps = runif(1,0,3),
                                   p = runif(1,0,1),
                                   z = zinits)
```

We specify the parameters to be monitored:

```r
parameters.to.save <- c("beta", "sdeps", "p")
```

We increase the number of iterations and the length of the burn-in
period to reach better convergence:

```r
n.iter <- 10000
n.burnin <- 5000
n.chains <- 2
```

And finally, we run NIMBLE:

```r
mcmc.phiwlrep <- nimbleMCMC(code = hmm.phiwlrep,
                            constants = my.constants,
                            data = my.data,
                            inits = initial.values,
                            monitors = parameters.to.save,
                            niter = n.iter,
                            nburnin = n.burnin,
                            nchains = n.chains)
```

We inspect the numerical summaries:

```r
MCMCsummary(mcmc.phiwlrep, round = 2)
```

```
##            mean    sd   2.5%    50% 97.5% Rhat n.eff
## beta[1]    0.22  0.12  -0.02   0.22  0.44 1.24  1335
## beta[2]   -0.01  0.10  -0.20  -0.01  0.19 1.02  1894
## p          0.90  0.03   0.84   0.90  0.95 1.03   752
## sdeps      0.27  0.25   0.01   0.16  0.83 4.68    12
```

The effective sample size for the standard deviation of the random effect is very low. Let's try something else, which means that the MCMC chains are exploring that parameter space inefficiently. A common strategy to address this problem is to reparameterize the model using a *non-centered* formulation.

In the *centered parameterization* we've used so far, the random effect is expressed directly in terms of its variance (or standard deviation). This often creates strong correlations between the random effects and their variance parameter, which in turn can slow down mixing and reduce the effective sample size. In the *non-centered parameterization*, we instead separate the structure of the random effect from its scale. In practice, we introduce a standardized random effect $\varepsilon_i \sim N(0, 1)$ and then multiply it by the standard deviation σ. This decouples the randomness from the scale parameter and usually improves sampling efficiency. In code, this looks like:

```
...
  for (i in 1:N){
    logit(phi[i])<-beta[1]+beta[2]*winglength[i] + sdeps * eps[i]
    eps[i] ~ dnorm(mean = 0, sd = 1)
...
```

After running NIMBLE, we inspect the numerical summaries, and see that effective sample sizes are much better:

```
MCMCsummary(mcmc.phiwlrep, round = 2)
```

```
##            mean   sd  2.5%    50% 97.5% Rhat n.eff
## beta[1]    0.21 0.12 -0.04   0.21  0.43 1.00  1202
## beta[2]   -0.01 0.10 -0.20  -0.01  0.19 1.00  1789
## p          0.90 0.03  0.83   0.90  0.95 1.01   790
## sdeps      0.40 0.26  0.02   0.37  0.95 1.01   170
```

4.8.5   Individual time-varying covariates

So far, we allowed covariates to vary along a single dimension, either time or individual. What if we need to consider a covariate that varies from one animal to the other, and over time. Think of age for example, its value is specific to each individual, and it (sadly) changes over time.

Age has a particular meaning in the capture-recapture framework. It is the time elapsed since first encounter, which is a proxy of true age, but not true age. If age is known at first encounter, then it is true age. For example, if the dippers were marked as young, then we would have the true biological age of each bird.

The convenient thing is that age has no missing value because age at $t + 1$ is just age at t to which we add 1. This suggests a way to code the age effect in NIMBLE as follows:

```
hmm.phiage.in <- nimbleCode({
  p ~ dunif(0, 1)                        # prior detection
  omega[1,1] <- 1 - p                    # Pr(alive t -> non-detected t)
```

```
  omega[1,2] <- p                          # Pr(alive t -> detected t)
  omega[2,1] <- 1                          # Pr(dead t -> non-detected t)
  omega[2,2] <- 0                          # Pr(dead t -> detected t)
  for (i in 1:N){
    for (t in first[i]:(T-1)){
    # phi1 = beta1 + beta2; phi1+ = beta1
    logit(phi[i,t]) <- beta[1] + beta[2] * equals(t, first[i])
    gamma[1,1,i,t] <- phi[i,t]             # Pr(alive t -> alive t+1)
    gamma[1,2,i,t] <- 1 - phi[i,t]         # Pr(alive t -> dead t+1)
    gamma[2,1,i,t] <- 0                    # Pr(dead t -> alive t+1)
    gamma[2,2,i,t] <- 1                    # Pr(dead t -> dead t+1)
    }
  }
  beta[1] ~ dnorm(mean = 0, sd = 1.5) # phi1+
  beta[2] ~ dnorm(mean = 0, sd = 1.5) # phi1 - phi1+
  phi1plus <- plogis(beta[1])             # phi1+
  phi1 <- plogis(beta[1] + beta[2])       # phi1
  delta[1] <- 1                           # Pr(alive t = 1) = 1
  delta[2] <- 0                           # Pr(dead t = 1) = 0
  # likelihood
  for (i in 1:N){
    z[i,first[i]] ~ dcat(delta[1:2])
    for (j in (first[i]+1):T){
      z[i,j] ~ dcat(gamma[z[i,j-1], 1:2, i, j-1])
      y[i,j] ~ dcat(omega[z[i,j], 1:2])
    }
  }
})
```

Here we used `equals(t, first[i])` which renders 1 when t is the first
encounter `first[i]` and 0 otherwise. Therefore, we distinguish survival over the first interval after first encounter ϕ_1 (`logit(phi[i,t]) <-`
`beta[1] + beta[2]`) from survival afterwards ϕ_{1+} (`logit(phi[i,t]) <-`
`beta[1]`).

We put all constants in a list:

```r
first <- apply(y, 1, function(x) min(which(x !=0)))
my.constants <- list(N = nrow(y),
                     T = ncol(y),
                     first = first)
```

And the data in a list:

```r
my.data <- list(y = y + 1)
```

We write a function to generate initial values:

```r
zinits <- y
zinits[zinits == 0] <- 1
initial.values <- function() list(beta = rnorm(2,0,5),
                                  p = runif(1,0,1),
                                  z = zinits)
```

And specify the parameters to be monitored:

```r
parameters.to.save <- c("phi1", "phi1plus", "p")
```

We now run NIMBLE:

```r
mcmc.phi.age.in <- nimbleMCMC(code = hmm.phiage.in,
                              constants = my.constants,
                              data = my.data,
                              inits = initial.values,
                              monitors = parameters.to.save,
                              niter = n.iter,
                              nburnin = n.burnin,
                              nchains = n.chains)
```

We display the results:

```
MCMCsummary(mcmc.phi.age.in, round = 2)
```

```
##             mean   sd 2.5%  50% 97.5% Rhat n.eff
## p          0.90 0.03 0.83 0.90  0.95 1.00   738
## phi1       0.56 0.03 0.49 0.56  0.62 1.00  1624
## phi1plus   0.57 0.04 0.49 0.57  0.64 1.01   506
```

Age or time elapsed since first encounter does not seem to have an effect on survival here.

Another method to include an age effect is to create an individual by time covariate and use nested indexing (as in the flood/nonflood example) to distinguish survival over the interval after first detection from survival afterwards:

```r
age <- matrix(NA, nrow = nrow(y), ncol = ncol(y) - 1)
for (i in 1:nrow(age)){
  for (j in 1:ncol(age)){
    if (j == first[i]) age[i,j] <- 1 # age = 1
    if (j > first[i]) age[i,j] <- 2  # age > 1
  }
}
```

Now we may write the NIMBLE code for this model. We just need to remember that survival is no longer defined on the logit scale as in the previous model, so we simply use uniform priors:

```r
hmm.phiage.out <- nimbleCode({
  p ~ dunif(0, 1)                  # prior detection
  omega[1,1] <- 1 - p              # Pr(alive t -> non-detected t)
  omega[1,2] <- p                  # Pr(alive t -> detected t)
  omega[2,1] <- 1                  # Pr(dead t -> non-detected t)
  omega[2,2] <- 0                  # Pr(dead t -> detected t)
  for (i in 1:N){
    for (t in first[i]:(T-1)){
```

```
    phi[i,t] <- beta[age[i,t]]      # beta1 = phi1, beta2 = phi1+
    gamma[1,1,i,t] <- phi[i,t]      # Pr(alive t -> alive t+1)
    gamma[1,2,i,t] <- 1 - phi[i,t]  # Pr(alive t -> dead t+1)
    gamma[2,1,i,t] <- 0             # Pr(dead t -> alive t+1)
    gamma[2,2,i,t] <- 1             # Pr(dead t -> dead t+1)
    }
  }
  beta[1] ~ dunif(0, 1) # phi1
  beta[2] ~ dunif(0, 1) # phi1+
  phi1 <- beta[1]
  phi1plus <- beta[2]
  delta[1] <- 1                            # Pr(alive t = 1) = 1
  delta[2] <- 0                            # Pr(dead t = 1) = 0
  # likelihood
  for (i in 1:N){
    z[i,first[i]] ~ dcat(delta[1:2])
    for (j in (first[i]+1):T){
      z[i,j] ~ dcat(gamma[z[i,j-1], 1:2, i, j-1])
      y[i,j] ~ dcat(omega[z[i,j], 1:2])
    }
  }
})
```

We put all constants in a list, including the age covariate:

```
first <- apply(y, 1, function(x) min(which(x !=0)))
my.constants <- list(N = nrow(y),
                     T = ncol(y),
                     first = first,
                     age = age)
```

We re-write a function to generate initial values:

```
zinits <- y
zinits[zinits == 0] <- 1
```

```r
initial.values <- function() list(beta = runif(2,0,1),
                                   p = runif(1,0,1),
                                   z = zinits)
```

And we run NIMBLE:

```r
mcmc.phi.age.out <- nimbleMCMC(code = hmm.phiage.out,
                               constants = my.constants,
                               data = my.data,
                               inits = initial.values,
                               monitors = parameters.to.save,
                               niter = n.iter,
                               nburnin = n.burnin,
                               nchains = n.chains)
```

We display numerical summaries for the model parameters, and acknowledge that we obtain similar results to the other parameterization:

```r
MCMCsummary(mcmc.phi.age.out, round = 2)
```

```
##           mean   sd 2.5%  50% 97.5% Rhat n.eff
## p         0.90 0.03 0.84 0.90  0.95 1.01   793
## phi1      0.55 0.03 0.48 0.55  0.62 1.01  1736
## phi1plus  0.57 0.04 0.50 0.57  0.64 1.00  2064
```

Like I mentioned earlier, age is easy to deal with as it does not contain missing values. Now think of size or body mass for a minute. The problem is that we cannot record size or body mass when an animal is non-detected. The easiest way to cope with individual time-varying covariates is to discretize e.g. in small, medium and large as in Chapter 5. Another option is to come up with a model for the covariate and fill in missing values by simulating from this model.

4.9   Why Bayes? Incorporate prior information

4.9.1   Prior elicitation

Before we close this section, I'd like to cover one last topic. Think again of the CJS model with constant parameters. So far, we have assumed a non-informative prior on survival $\text{Beta}(1, 1) = \text{Uniform}(0, 1)$. With this prior, we have seen in Section 4.5 that mean posterior survival is $\phi = 0.56$ with credible interval $[0.52, 0.62]$.

The thing is that we know a lot about passerines and it is a shame not to be able to use this information and act as if we have to start from scratch and know nothing.

We illustrate how to incorporate prior information by acknowledging that species with similar body masses have similar survival. By gathering information on several other European passerines than the dipper, let's assume we have built a regression of survival vs. body mass – an allometric relationship.

Now knowing dippers weigh on average 59.8g, we're in the position to build a prior for dipper survival probability by predicting its value using the regression. We obtain a predicted survival of 0.57 and a standard deviation of 0.075. Using an informative prior `phi ~ dnorm(0.57, sd = 0.073)` in NIMBLE instead of our `phi ~ dunif(0,1)`, we get a mean posterior of 0.56 with credible interval $[0.52, 0.61]$. There's barely no difference with the non-informative prior, quite a disappointment.

Now let's assume that we had only the three first years of data, what would have happened? We fit the model with constant parameters with both the non-informative and informative priors to the dataset from which we delete the final 4 years of data. Now the benefit of using the prior information becomes clear as the credible interval when prior information is ignored has a width of 0.53, which is more than twice as much as when prior information is used (0.24), illustrating the increased precision provided by the prior. We may assess visually this gain in precision by comparing the survival posterior distributions with and without informative prior:

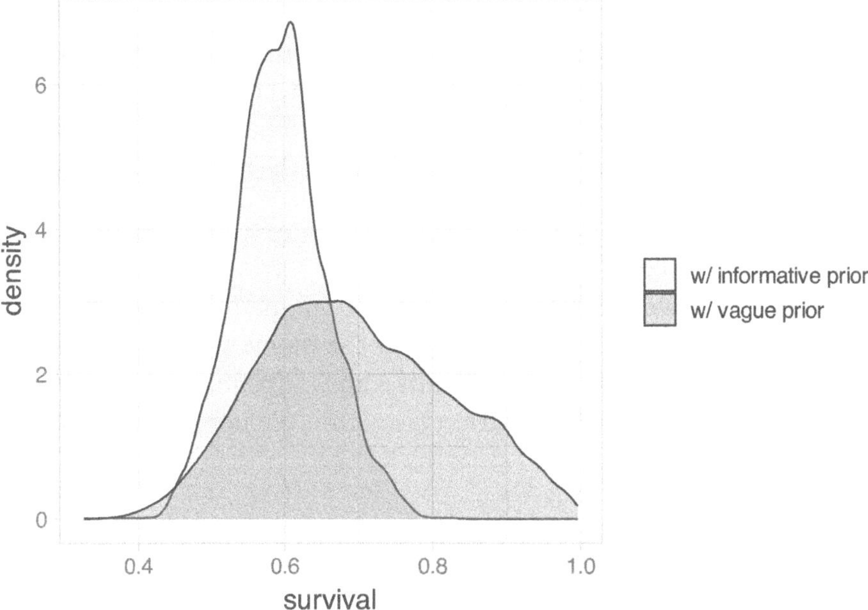

If the aim is to get an estimate of survival, Gilbert did not have to conduct further data collection after 3 years, and he could have reached the same precision as with 7 years of data by using prior information derived from body mass. In brief, the prior information was worth 4 years of field data. Of course, this is assuming that the ecological question remains the same whether you have 3 or 7 years of data, which is unlikely to be the case, as with long-term data, there is so much we can ask, more than "just" what annual survival probability is.

4.9.2 Moment matching

The prior `phi ~ dnorm(0.57, sd = 0.073)` is not entirely satisfying because it is not constrained to be positive or less than one, which is the minimum for a probability (of survival) to be well defined. In our specific example, the prior distribution is centered on positive values far from 0, and the standard deviation is small enough so that the chances to get values smaller than 0 or higher than 1 are null (to convince yourself, just type in `hist(rnorm(1000, mean = 0.57, sd = 0.073))` in R). Can we do better? The answer is yes.

Remember the Beta distribution? Recall that the Beta distribution is a continuous distribution with values between 0 and 1, which is very convenient to specify priors for survival and detection probabilities. Besides we know everything about the Beta distribution, in particular its mean and variance. If $X \sim Beta(\alpha, \beta)$, then the mean and (variance) of X are $\mu = \mathrm{E}(X) = \frac{\alpha}{\alpha+\beta}$ and $\sigma^2 = \mathrm{Var}(X) = \frac{\alpha\beta}{(\alpha+\beta)^2(\alpha+\beta+1)}$.

In the capture-recapture example, we know a priori that the mean of the probability we're interested in is $\mu = 0.57$ and its variance is $\sigma^2 = 0.073^2$. Parameters μ and σ^2 are seen as the moments of a $Beta(\alpha, \beta)$ distribution. Now we look for values of α and β that match the observed moments of the Beta distribution μ and σ^2. We need another set of equations:

$$\alpha = \left(\frac{1-\mu}{\sigma^2} - \frac{1}{\mu} \right) \mu^2$$

$$\beta = \alpha \left(\frac{1}{\mu} - 1 \right)$$

For our model, that means:

```
(alpha <- ( (1 - 0.57)/(0.073*0.073) - (1/0.57) )*0.57^2)
## [1] 25.65
(beta <- alpha * ( (1/0.57) - 1))
## [1] 19.35
```

Now we simply have to use `phi ~ dbeta(25.6,19.3)` as a prior instead of our `phi ~ dnorm(0.57, sd = 0.073)`.

4.10   Summary

- The CJS model is a HMM with time-varying parameters to account for variation due to environmental conditions in survival or to sampling effort in detection.

- Covariates can be considered to try and explain temporal and/or individual variation in survival and detection probabilities. If needed, random effects can be added to cope with unexplained variation.

- For model comparison, the WAIC can be used to evaluate the relative predictive performance of capture-recapture models.

- Statistical models rely on assumptions, and the CJS model makes no exception. There are procedures to assess the goodness of fit of the CJS model to capture-recapture data.

4.11   Suggested reading

- Buckland [2016] provides a historical perspective on the CJS model and anecdotes. The monography by Lebreton et al. [1992] is a must-read to better understand the CJS model and its applications. The HMM formulation of the CJS model was proposed by Gimenez et al. [2007] and Royle [2008].

- Gimenez et al. [2009b] deals with parameter redundancy in capture-recapture models in a Bayesian framework. For an exhaustive treatment, see Cole [2020] excellent book.

- Relative to model comparison, I warmly recommend McElreath [2020] to better understand WAIC, with the accompanying video, see https://www.youtube.com/watch?v=vSjL2Zc-gEQ. The paper by Gelman et al. [2014] is also much helpful.

- On posterior predictive checks, you may check out Conn et al. [2018]. The `R2ucare` package is introduced by Gimenez et al. [2018b].

- Temporal heterogeneity is addressed in papers by Grosbois et al. [2008] and Frederiksen et al. [2014], while individual heterogeneity is reviewed by Gimenez et al. [2018a].

- Regarding covariates, I did not use the formalization of linear models and sticked to an intuitive (hopefully) illustration of the use of covariates. More details can be found in Chapter 6 of Cooch and White [2017].

- The example on how to incorporate prior information is from McCarthy and Masters [2005].

5

Sites and states

5.1 Introduction

In this fifth chapter, you will learn about the Arnason-Schwarz model that allows estimating transitions between sites and states based on capture-recapture data. You will also see how to deal with uncertainty in the assignment of states to individuals.

5.2 The Arnason-Schwarz (AS) model

In Chapter 4, we got acquainted with the Cormack-Jolly-Seber (CJS) model which accommodates transitions between the states alive and dead while accounting for imperfect detection. It is often the case that besides being alive, more detailed information is collected on the state of animals when they are detected. For example, if the study area is split into several discrete sites, you may record where an animal is detected, the state being now alive in this particular site. The Arnason-Schwarz (AS) model can be viewed as an extension to the CJS model in which we estimate movements between sites on top of survival. The AS model is named after the two statisticians – Neil Arnason and Carl Schwarz – who came up with the idea.

5.2.1 Theory

Let's assume for now that we have two sites, say 1 and 2. The way we usually think of analyzing the data is to start from the detections and non-detections and infer the transitions between sites and movements. Schematically, when an animal is detected in site 1 or site 2, it obviously

DOI: 10.1201/9781003244097-5

means that it is alive in that site, whereas when it is not detected, it may be dead or alive in either site. Schematically, we have:

Observations States

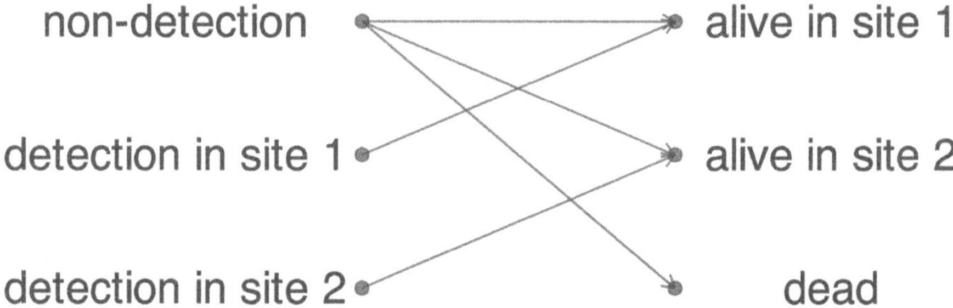

Observations and states are indeed closely related, but we do not have a perfect states to observations correspondence, and the HMM framework will help you make the distinction clear, which in turn will make the modeling easier. Usually, we focus our energy on the observations but what we'd really like is to spend time thinking of the ecological processes that we observed imperfectly. In the HMM framework, when we are to build a model, we think of the states and their dynamic over time, and these states emit the observations we're given to make. Going back to the our example, when an animal is alive in either site, it may get detected in that site or go undetected. When an animal is dead, then it goes undetected for sure. Schematically, we obtain:

States Observations

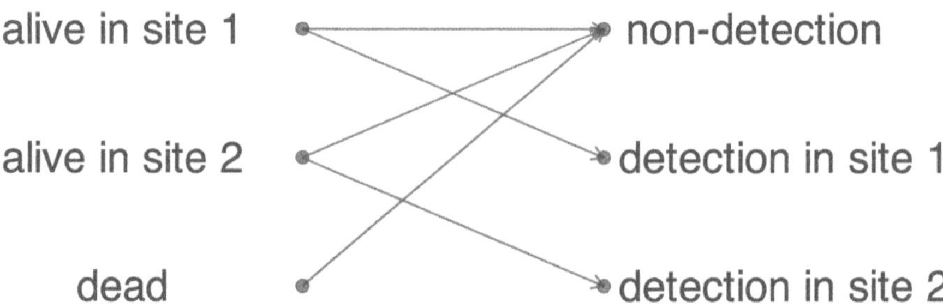

We have $z = 1$ for alive in site 1, $z = 2$ for alive in site 2 and $z = 3$ for dead. We will code $y = 1$ for non-detected, $y = 2$ for detected in site 1 and $y = 3$ for detected in site 2. The parameters are:
- π^r is the probability the a newly encountered individual is in state r;
- p_t^r is the probability of detection at t for a bird in site r at t;
- ϕ_t^r is the survival probability for birds in site r at between t and $t + 1$;
- ψ_t^{rs} is the probability of being in site s at time $t+1$ for animals that were in site r at t and have survived to $t + 1$, in short movement conditional on survival.

These parameters can be modeled as functions of covariates as in Section 4.8. But for now we will drop the time index for simplicity.

We follow the presentation of HMM as in Chapter 3. Let's start with the vector of initial states. At first encounter, an animal may be alive in site 1 with probability π^1 or in site 2 with the complementary probability $1 - \pi^1$, but it cannot be dead. Therefore, we have:

$$\begin{array}{ccc} z_t = 1 & z_t = 2 & z_t = 3 \end{array}$$
$$\delta = \begin{pmatrix} \pi^1 & 1 - \pi^1 & 0 \end{pmatrix}$$

We move on to the transition matrix which connects the states at $t - 1$ in rows to the states at t in columns. For example, the probability of moving from site 1 at $t - 1$ to site 2 at t is the product of the survival probability in site 1 over that time interval, times the probability of moving from site 1 to 2 for those animals who survived. We have:

$$\begin{array}{ccc} z_t = 1 & z_t = 2 & z_t = 3 \end{array}$$
$$\Gamma = \begin{pmatrix} \phi^1(1 - \psi^{12}) & \phi^1\psi^{12} & 1 - \phi^1 \\ \phi^2\psi^{21} & \phi^2(1 - \psi^{21}) & 1 - \phi^2 \\ 0 & 0 & 1 \end{pmatrix} \begin{array}{c} z_{t-1} = 1 \\ z_{t-1} = 2 \\ z_{t-1} = 3 \end{array}$$

Finally, the observation matrix relates the states an animal is in at t in rows to the observations at t in columns. If an animal is dead ($z_t = 3$), it cannot be detected ($\Pr(y_t = 1|z_t = 3) = \Pr(y_t = 2|z_t = 3) = 0$ and $\Pr(y_t = 3|z_t = 3) = 1$), whereas when it is alive in either site, it can be detected or not. We have:

$$\Omega = \begin{array}{ccc} y_t = 1 & y_t = 2 & y_t = 3 \end{array}$$

$$\Omega = \begin{pmatrix} 1 - p^1 & p^1 & 0 \\ 1 - p^2 & 0 & p^2 \\ 1 & 0 & 0 \end{pmatrix} \begin{array}{l} z_t = 1 \\ z_t = 2 \\ z_t = 3 \end{array}$$

The vector of initial state probabilities sums up to one, as well as the rows of the transition and observation matrices.

5.3   Fitting the AS model

5.3.1   Geese data

To introduce this chapter, we will use data on the Canada goose (*Branta canadensis*; geese hereafter) kindly provided by Jay Hestbeck (Figure 4.2).

FIGURE 5.1: Canada goose (*Branta canadensis*). Credit: Max McCarthy.

In total, 21277 geese were captured, marked with coded neck bands and recaptured between 1984 and 1989 in their wintering locations. Specifically, geese were monitored in the Atlantic flyway, in large areas along the East coast of the USA, namely 3 sites in the mid–Atlantic (New York, Pennsylvania, New Jersey), Chesapeake (Delaware, Maryland, Virginia), and Carolinas (North and South Carolina). Birds were adults and sub-adults when banded.

You may see the data below:

```
y <- read_csv("geese.csv") %>% as.matrix()
head(y)
```

```
##      year_1984 year_1985 year_1986 year_1987 year_1988
## [1,]         0         2         2         0         0
## [2,]         0         0         0         0         0
## [3,]         0         0         0         1         0
## [4,]         0         0         2         0         0
## [5,]         0         3         0         0         3
## [6,]         0         0         0         2         0
##      year_1989
## [1,]         0
## [2,]         2
## [3,]         0
## [4,]         0
## [5,]         2
## [6,]         0
```

The six columns are years in which the geese were captured, banded and recapture. A 0 stands for a non-detection, and detections were coded in the 3 wintering sites 1, 2 and 3 for mid–Atlantic, Chesapeake and Carolinas respectively. This is only a subsample of 500 individuals of the whole dataset that I will use for illustration here.

5.3.2   NIMBLE implementation

To write the NIMBLE code corresponding to the AS model, we will make our life easier and start with 2 sites only – we drop the Carolinas wintering site for now. We replace all 3's by 0's in the dataset:

```r
y[y==3] <- 0
# remove rows with 0's
mask <- apply(y, 1, sum)
y <- y[mask!=0,]
```

Also we consider parameters constant over time.

We start with comments to define the quantities – parameters, states and observations – we will use in the NIMBLE code:

```r
multisite <- nimbleCode({
  # ------------------------------------------------------------
  # Parameters:
  # phi1: survival probability site 1
  # phi2: survival probability site 2
  # psi12: movement probability from site 1 to site 2
  # psi21: movement probability from site 2 to site 1
  # p1: detection probability site 1
  # p2: detection probability site 2
  # ------------------------------------------------------------
  # States (z):
  # 1 alive at site 1
  # 2 alive at site 2
  # 3 dead
  # Observations (y):
  # 1 not seen
  # 2 seen at site 1
  # 3 seen at site 2
  # ------------------------------------------------------------
...
```

The we specify priors for the survival, transition and detection probabilities:

```
multisite <- nimbleCode({

...

  # Priors
  phi1 ~ dunif(0, 1)
  phi2 ~ dunif(0, 1)
  psi12 ~ dunif(0, 1)
  psi21 ~ dunif(0, 1)
  p1 ~ dunif(0, 1)
  p2 ~ dunif(0, 1)

...
```

We now write the vector of initial state probabilities:

```
multisite <- nimbleCode({

...

  # initial state probabilities
  delta[1] <- pi1           # Pr(alive in 1 at t = first)
  delta[2] <- 1 - pi1       # Pr(alive in 2 at t = first)
  delta[3] <- 0             # Pr(dead at t = first) = 0

...
```

Actually, the initial state is known exactly: It is alive at site of initial capture, and π^1 is the proportion of individuals first captured in site 1, there is no need to make it explicit in the model and estimate it. Therefore, in the likelihood, instead of `z[i,first[i]] ~ dcat(delta[1:3])`, you can use `z[i,first[i]] <- y[i,first[i]] - 1` and just forget about π^1, for now. Note that the same trick applies to the CJS model.

We write the transition matrix:

```
multisite <- nimbleCode({

...

  # probabilities of state z(t+1) given z(t)
  # (read as gamma[z(t),z(t+1)] = gamma[fromState,toState])
```

```
gamma[1,1] <- phi1 * (1 - psi12)
gamma[1,2] <- phi1 * psi12
gamma[1,3] <- 1 - phi1
gamma[2,1] <- phi2 * psi21
gamma[2,2] <- phi2 * (1 - psi21)
gamma[2,3] <- 1 - phi2
gamma[3,1] <- 0
gamma[3,2] <- 0
gamma[3,3] <- 1
...
```

In the same way, the observation matrix is:

```
multisite <- nimbleCode({
...
  # probabilities of y(t) given z(t)
  # (read as omega[y(t),z(t)] = omega[Observation,State])

  omega[1,1] <- 1 - p1      # Pr(alive 1 t -> non-detected t)
  omega[1,2] <- p1          # Pr(alive 1 t -> detected site 1 t)
  omega[1,3] <- 0           # Pr(alive 1 t -> detected site 2 t)
  omega[2,1] <- 1 - p2      # Pr(alive 2 t -> non-detected t)
  omega[2,2] <- 0           # Pr(alive 2 t -> detected site 1 t)
  omega[2,3] <- p2          # Pr(alive 2 t -> detected site 2 t)
  omega[3,1] <- 1           # Pr(dead t -> non-detected t)
  omega[3,2] <- 0           # Pr(dead t -> detected site 1 t)
  omega[3,3] <- 0           # Pr(dead t -> detected site 2 t)
...
```

At last, we are ready to specify the likelihood which, and this this the
magic of HMM, is the same as the likelihood of the CJS model, only the
vector of initial states, the transition and observation matrices were
changed:

```
multisite <- nimbleCode({
...
  # likelihood
```

```r
  for (i in 1:N){
    # latent state at first capture
    z[i,first[i]] <- y[i,first[i]] - 1
    for (t in (first[i]+1):K){
      # z(t) given z(t-1)
      z[i,t] ~ dcat(gamma[z[i,t-1],1:3])
      # y(t) given z(t)
      y[i,t] ~ dcat(omega[z[i,t],1:3])
    }
  }
})
```

We need to provide NIMBLE with constants, data, initial values, some parameters to monitor and MCMC details:

```r
# occasions of first capture
first <- apply(y, 1, function(x) min(which(x !=0)))
# constants
my.constants <- list(first = first,
                     K = ncol(y),
                     N = nrow(y))
# data
my.data <- list(y = y + 1)
# initial values
zinits <- y # say states are observations, detections in A or B
           # are taken as alive in same sites
# non-detections become alive in site A or B
zinits[zinits==0] <- sample(c(1,2), sum(zinits==0), replace = TRUE)
initial.values <- function(){list(phi1 = runif(1, 0, 1),
                                  phi2 = runif(1, 0, 1),
                                  psi12 = runif(1, 0, 1),
                                  psi21 = runif(1, 0, 1),
                                  p1 = runif(1, 0, 1),
                                  p2 = runif(1, 0, 1),
                                  pi1 = runif(1, 0, 1),
                                  z = zinits)}
```

```
# parameters to monitor
parameters.to.save <- c("phi1", "phi2","psi12", "psi21",
                        "p1", "p2", "pi1")
# MCMC details
n.iter <- 20000
n.burnin <- 1000
n.chains <- 2
```

Now we may run NIMBLE:

```
mcmc.multisite <- nimbleMCMC(code = multisite,
                             constants = my.constants,
                             data = my.data,
                             inits = initial.values,
                             monitors = parameters.to.save,
                             niter = n.iter,
                             nburnin = n.burnin,
                             nchains = n.chains)
```

We may have a look to the results via a caterpillar plot:

```
MCMCplot(mcmc.multisite)
```

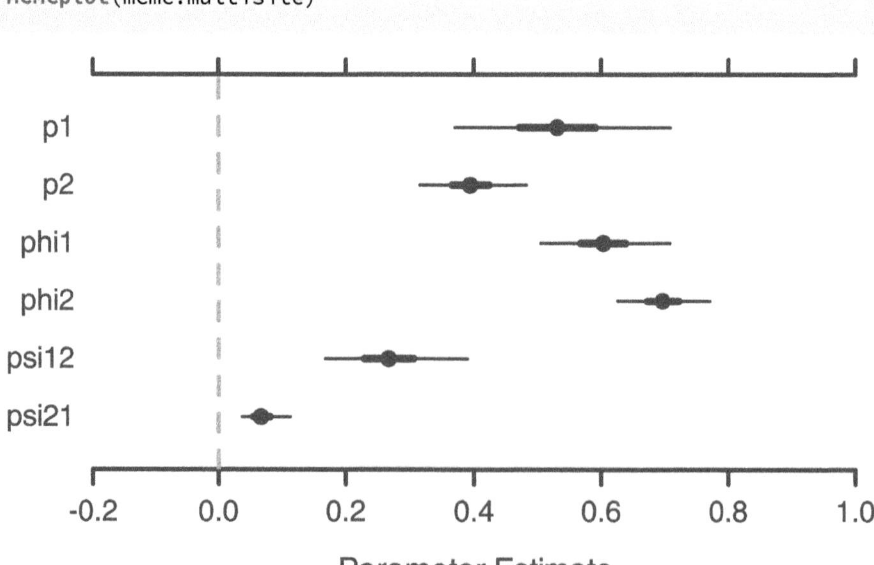

Remember mid–Atlantic is site 1, and Chesapeake site 2. Detection in mid–Atlantic (around 0.5) is higher than in Chesapeake (around 0.4) although it comes with more uncertainty (wider credible interval). Survival in both sites are estimated at around 0.6–0.7. Note that by going multisite, we could make these parameters site-specific and differences might reflect habitat quality for example. Now the novelty lies in our capability to estimate movements from site 1 to site 2 and from site 2 to site 1 from a winter to the next. The annual probability of remaining in the same site for two successive winters, used as a measure of site fidelity, was lower in the mid–Atlantic ($1 - \psi_{12}$ around 0.8) than in the Chesapeake ($1 - \psi_{21}$ around 0.9). The estimated probability of moving to the Chesapeake from the mid–Atlantic was four times as high as the probability of moving in the opposite direction.

We may also have a look to numerical summaries, which confirm our ecological interpretation of the model parameter estimates:

```
MCMCsummary(mcmc.multisite, round = 2)
##          mean   sd 2.5%  50% 97.5% Rhat n.eff
## p1       0.53 0.09 0.37 0.53  0.71    1  1107
## p2       0.40 0.04 0.32 0.39  0.48    1  1080
## phi1     0.60 0.05 0.50 0.60  0.71    1  1675
## phi2     0.70 0.04 0.63 0.70  0.77    1  1145
## psi12    0.27 0.06 0.17 0.27  0.39    1  2001
## psi21    0.07 0.02 0.04 0.07  0.11    1  2192
```

You could add time-dependence to the demographic parameters, e.g. survival and movements, and assess the effect of winter harshness with some temporal covariates; Individual covariates could also be considered. See Section 4.8.

Before we move to the next section, I will illustrate the use of `nimbleEcology` to fit the AS to the geese data with 2 sites (see Section 3.8.3). Using the function `dHMM()` which implements HMM with time-independent observation and transition matrices, we have:

```r
# read in data
geese <- read_csv("geese.csv", col_names = TRUE)
y <- as.matrix(geese)
# drop Carolinas
y[y==3] <- 0 # act as if there was no detection in site 3 Carolinas
mask <- apply(y, 1, sum)
y <- y[mask!=0,] # remove rows w/ 0s only
# get occasions of first encounter
get.first <- function(x) min(which(x != 0))
first <- apply(y, 1, get.first)
# filter out individuals that are first captured at last occasion.
# These individuals do not contribute to parameter estimation,
# and also they cause problems with nimbleEcology
mask <- which(first!=ncol(y))
y <- y[mask, ]                    # keep only these
first <- first[mask]


# NIMBLE code
multisite.marginalized <- nimbleCode({


  # -----------------------------------------------------
  # Parameters:
  # phi1: survival probability site 1
  # phi2: survival probability site 2
  # psi12: movement probability from site 1 to site 2
  # psi21: movement probability from site 2 to site 1
  # p1: recapture probability site 1
  # p2: recapture probability site 2
  # pi1: prop of being in site 1 at initial capture
  # -----------------------------------------------------
  # States (z):
  # 1 alive at 1
  # 2 alive at 2
  # 3 dead
  # Observations (y):
  # 1 not seen
  # 2 seen at site 1
  # 3 seen at site 2
```

```r
# ------------------------------------------------------------

# priors
phi1 ~ dunif(0, 1)
phi2 ~ dunif(0, 1)
psi12 ~ dunif(0, 1)
psi21 ~ dunif(0, 1)
p1 ~ dunif(0, 1)
p2 ~ dunif(0, 1)
pi1 ~ dunif(0, 1)

# probabilities of state z(t+1) given z(t)
gamma[1,1] <- phi1 * (1 - psi12)
gamma[1,2] <- phi1 * psi12
gamma[1,3] <- 1 - phi1
gamma[2,1] <- phi2 * psi21
gamma[2,2] <- phi2 * (1 - psi21)
gamma[2,3] <- 1 - phi2
gamma[3,1] <- 0
gamma[3,2] <- 0
gamma[3,3] <- 1

# probabilities of y(t) given z(t)
omega[1,1] <- 1 - p1      # Pr(alive 1 t -> non-detected t)
omega[1,2] <- p1          # Pr(alive 1 t -> detected 1 t)
omega[1,3] <- 0           # Pr(alive 1 t -> detected 2 t)
omega[2,1] <- 1 - p2      # Pr(alive 2 t -> non-detected t)
omega[2,2] <- 0           # Pr(alive 2 t -> detected 1 t)
omega[2,3] <- p2          # Pr(alive 2 t -> detected 2 t)
omega[3,1] <- 1           # Pr(dead t -> non-detected t)
omega[3,2] <- 0           # Pr(dead t -> detected 1 t)
omega[3,3] <- 0           # Pr(dead t -> detected 2 t)

# likelihood
# initial state probs
for(i in 1:N) {
  # first state propagation
  init[i, 1:3] <- gamma[ y[i, first[i] ] - 1, 1:3 ]
```

```r
  }
  for (i in 1:N){
    y[i,(first[i]+1):K] ~ dHMM(init = init[i,1:3],# data first[i]+1
                               probObs = omega[1:3,1:3], # obs
                               probTrans = gamma[1:3,1:3], # trans
                               len = K - first[i], # nb of occasions
                               checkRowSums = 0) # do not check
                                                 # whether elements
                                                 # in a row sum up
                                                 # to 1
  }
})
# constants
my.constants <- list(first = first,
                     K = ncol(y),
                     N = nrow(y))
# data
my.data <- list(y = y + 1)
# initial values
initial.values <- function(){list(phi1 = runif(1, 0, 1),
                                   phi2 = runif(1, 0, 1),
                                   psi12 = runif(1, 0, 1),
                                   psi21 = runif(1, 0, 1),
                                   p1 = runif(1, 0, 1),
                                   p2 = runif(1, 0, 1))}
# parameters to monitor
parameters.to.save <- c("phi1", "phi2","psi12", "psi21", "p1", "p2")
# MCMC details
n.iter <- 5000
n.burnin <- 1000
n.chains <- 2
# run NIMBLE
mcmc.multisite.marginalized <- nimbleMCMC(
  code = multisite.marginalized,
  constants = my.constants,
  data = my.data,
  inits = initial.values,
  monitors = parameters.to.save,
```

```
  niter = n.iter,
  nburnin = n.burnin,
  nchains = n.chains)
```

We obtain:

```
MCMCsummary(mcmc.multisite.marginalized, round = 2)
##          mean    sd 2.5%   50% 97.5% Rhat  n.eff
## p1       0.53  0.09 0.36  0.53  0.71    1    651
## p2       0.40  0.04 0.32  0.39  0.49    1    618
## phi1     0.60  0.05 0.51  0.60  0.71    1    935
## phi2     0.70  0.04 0.63  0.70  0.76    1    804
## psi12    0.27  0.06 0.17  0.26  0.39    1    973
## psi21    0.07  0.02 0.04  0.07  0.11    1    965
```

There are no differences whatsoever in these parameters estimates compared to those we obtained earlier.

5.3.3 Goodness of fit

Like for the CJS model, we need to ensure that the AS model provides a good fit to the data. With regard to classical tests, the goodness-of-fit testing procedures we covered in Section 4.7 for the CJS model have been extended to the AS model. These procedure are also available in the package R2ucare. While the transience and trap-dependence tests can be used on each site, it is worth mentioning a test that is specific to the AS model with several sites. The "where before where after" (WBWA) tests for memory, in other words it's a test for the Markovian property of the HMM in the particular context of capture-recapture. WBWA quantifies whether there is a relationship between where an animal has last been seen and where it will next be seen again. If a relationship exists, positive or negative, this suggests memory in the movements, and the probability for an animal of moving to a site at t, given it is present in a site at $t - 1$ should be made dependent of where it was at $t - 2$. This is called a second-order Markov process.

Going back to the geese data, the WBWA test is implemented as follows on the whole dataset:

```r
# To install the R2ucare package:
# if(!require(devtools)) install.packages("devtools")
# devtools::install_github("oliviergimenez/R2ucare")
library(R2ucare)
geese <- read_csv2("allgeese.csv") %>% as.matrix()
y <- geese[,1:6]
size <- geese[,7]
wbwa <- test3Gwbwa(y, size)
wbwa$test3Gwbwa
```

```
##  stat     df p_val
## 472.9  20.0   0.0
```

There is clearly a strong (not to say significant) positive relationship judging by the value of the statistic. I will demonstrate how to account for this memory issue in an extension of the AS model in a case study in Section 7.3.

5.4   What if more than two sites?

So far we have considered two sites only, for the sake of simplicity. Indeed when going for three sites (or more), a difficulty arises. While the movement probabilities still need to be between 0 and 1, the sum of all probabilities of moving from a site should also sum to one. This is because an individual alive in site 1 for example, has to stay in 1, move to 2 or move to 3, it has no other choice. This is no problem when you have only two sites because the probability of moving from 1 to 2 is always estimated between 0 and 1, and its complementary, the probability of moving from 2 to 1 will be too. When you have three sites, it might happen that the sum of the estimates for ψ^{12} and ψ^{13} is larger than one, even though they're both between 0 and 1, which would make $\psi^{11} = 1 - \psi^{12} - \psi^{13}$ negative.

There are basically two methods to fulfill both constraints, either to assign a Dirichlet (which is pronounced deer-eesh-lay) prior to the movement probabilities, or to use a multinomial logit link function.

5.4.1 Dirichlet prior

The Dirichlet distribution extends the Beta distribution multivariate we have seen previously (Figure 1.1) for values between 0 and 1 that add up to 1. Going back to our example with 3 sites, we would like to come up with a prior for parameters of moving from 1, which are ψ^{11}, ψ^{12} and ψ^{13}. The Dirichlet distribution of dimension 3 has a vector of parameters $(\alpha_1, \alpha_2, \alpha_3)$ that controls its shape. The sum of all α's can be interpreted as a measure of precision: The higher the sum, the more peaked is the distribution on the mean value, the mean value along each dimension being the ratio of the corresponding over the sum of all α's. When all α's are equal, the distribution is symmetric (Figure 5.2). Its mean is $(1/3, 1/3, 1/3)$ in our example with 3 parameters, whatever the value of α. When $\alpha_1 = \alpha_2 = \alpha_3 = 1$, we obtain a uniform marginal distribution for the ψ's, while values below 1 ($\alpha_1 = \alpha_2 = \alpha_3 = 0.1$) result in the distribution concentrating in the corners (skewed U-shaped forms) and values above 1 ($\alpha_1 = \alpha_2 = \alpha_3 = 10$) result in unimodal marginal distributions.

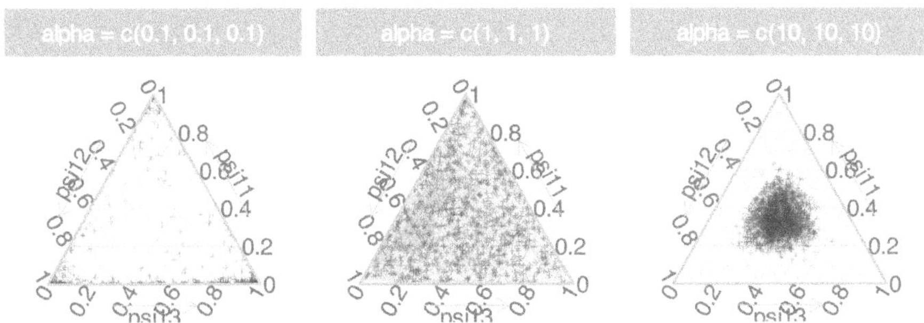

FIGURE 5.2: The Dirichlet distribution as a prior for $(\psi^{11}, \psi^{12}, \psi^{13})$ with vector of parameters α. Here all components of α are equal which makes the distribution symmetric, and its mean is $(1/3, 1/3, 1/3)$. When $\alpha = 1$, the prior for the ψ's is uniform (middle panel), unimodal when $\alpha = 10$ (right panel) and concentrated in the corners (0 and 1) when $\alpha = 0.1$ (left panel).

Going back to our example, in NIMBLE, we consider a Dirichlet prior for each triplet of movement parameters, from site 1 (ψ^{11}, ψ^{12} and ψ^{13}), from site 2 (ψ^{21}, ψ^{22} and ψ^{23}) and from site 3 (ψ^{31}, ψ^{32} and ψ^{33}).

We start by setting the scene with comments:

```
...
# ------------------------------------------------------------
# Parameters:
# phi1: survival prob site 1
# phi2: survival prob site 2
# phi3: survival prob site 3
# psi11 = psi1[1]: mov prob site 1 -> site 1 (reference)
# psi12 = psi1[2]: mov prob site 1 -> site 2
# psi13 = psi1[3]: mov prob site 1 -> site 3
# psi21 = psi2[1]: mov prob site 2 -> site 1
# psi22 = psi2[2]: mov prob site 2 -> site 2 (reference)
# psi23 = psi2[3]: mov prob site 2 -> site 3
# psi31 = psi3[1]: mov prob site 3 -> site 1
# psi32 = psi3[2]: mov prob site 3 -> site 2
# psi33 = psi3[3]: mov prob site 3 -> site 3 (reference)
# p1: recapture prob site 1
# p2: recapture prob site 2
# p3: recapture prob site 3
# ------------------------------------------------------------
# States (z):
# 1 alive at 1
# 2 alive at 2
# 2 alive at 3
# 3 dead
# Observations (y):
# 1 not seen
# 2 seen at 1
# 3 seen at 2
# 3 seen at 3
...
```

```
multisite <- nimbleCode({
...
  # transitions: Dirichlet priors
  psi1[1:3] ~ ddirch(alpha[1:3]) # psi11, psi12, psi13
  psi2[1:3] ~ ddirch(alpha[1:3]) # psi21, psi22, psi23
  psi3[1:3] ~ ddirch(alpha[1:3]) # psi31, psi32, psi33
...
```

Then we use these parameters (which now respect the constraints) to define the transition matrix:

```
multisite <- nimbleCode({
...
  # probabilities of state z(t+1) given z(t)
  gamma[1,1] <- phi1 * psi1[1]
  gamma[1,2] <- phi1 * psi1[2]
  gamma[1,3] <- phi1 * psi1[3]
  gamma[1,4] <- 1 - phi1
  gamma[2,1] <- phi2 * psi2[1]
  gamma[2,2] <- phi2 * psi2[2]
  gamma[2,3] <- phi2 * psi2[3]
  gamma[2,4] <- 1 - phi2
  gamma[3,1] <- phi3 * psi3[1]
  gamma[3,2] <- phi3 * psi3[2]
  gamma[3,3] <- phi3 * psi3[3]
  gamma[3,4] <- 1 - phi3
  gamma[4,1] <- 0
  gamma[4,2] <- 0
  gamma[4,3] <- 0
  gamma[4,4] <- 1
...
```

When we fit this model to the geese dataset with the detections in the Carolinas wintering site back in, and with `alpha <- c(1, 1, 1)` passed to the constants, we obtain the following results:

```
MCMCsummary(mcmc.multisite, round = 2)
##             mean    sd 2.5%  50% 97.5% Rhat n.eff
## p1          0.53 0.09 0.36 0.52  0.70 1.01   412
## p2          0.46 0.05 0.37 0.45  0.57 1.00   399
## p3          0.24 0.06 0.13 0.23  0.38 1.01   248
## phi1        0.60 0.05 0.50 0.60  0.70 1.01   564
## phi2        0.70 0.04 0.63 0.70  0.77 1.00   547
## phi3        0.78 0.07 0.64 0.78  0.91 1.00   258
## psi1[1]     0.74 0.06 0.62 0.74  0.84 1.00   798
## psi1[2]     0.24 0.05 0.14 0.23  0.36 1.00   853
## psi1[3]     0.02 0.03 0.00 0.02  0.10 1.01   401
## psi2[1]     0.07 0.02 0.04 0.07  0.12 1.01   687
## psi2[2]     0.83 0.04 0.73 0.84  0.90 1.00   363
## psi2[3]     0.09 0.04 0.04 0.09  0.18 1.01   361
## psi3[1]     0.02 0.02 0.00 0.02  0.06 1.00  1631
## psi3[2]     0.21 0.05 0.12 0.20  0.32 1.00   721
## psi3[3]     0.77 0.06 0.65 0.78  0.87 1.00   672
```

Survival probabilities are similar among sites, although lower in the mid-Atlantic (`phi[1]`). The detection probability in Carolinas (`p3`) seems much lower than in the two other wintering sites. The estimated probability of moving to the Chesapeake from the Carolinas (`psi3[2]`) is 2 times as high as the probability of moving in the opposite direction (`psi2[3]`).

In theory, you could include covariates as in Section 4.8 through the α parameters and the use a log link function (to ensure $\alpha > 0$), e.g. $\log(\alpha) = \beta_1 + \beta_2 x$. However, NIMBLE does not allow that. Fortunately, there is another way to specify the Dirichlet distribution through a ratio of gamma distributions that allows to incorporate covariates.

The gamma distribution is continuous. It has two parameters α and θ that control its shape and scale (Figure 5.3). Another parameterization considers its shape and rate which is the inverse of the scale.

If we have three independent random variables Y_1, Y_2 and Y_3 distributed as gamma distributions with shape parameters the α's and same scale parameter θ, that is $Y_j \sim \text{Gamma}(\alpha_j, \theta)$, then it can be

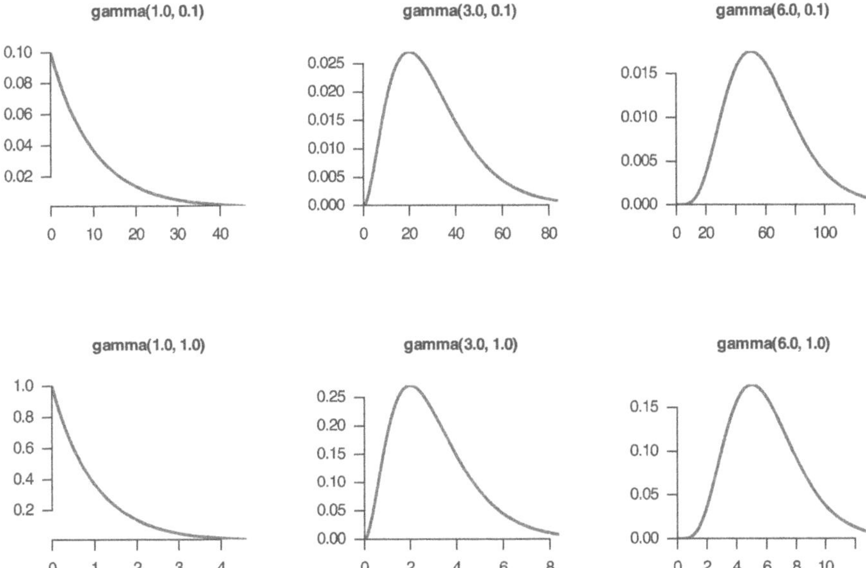

FIGURE 5.3: The distribution gamma(α,θ) for different values of α and θ. The shape argument α determines the overall shape while the scale parameter θ only affects the scale values (compare the values on X- and Y-axes between the bottom and upper panels). The exponential and chi-square distributions are particular cases of the gamma distribution. If the parameter shape is close to zero, the gamma is very similar to the exponential (bottom and upper left panels). If the parameter shape is large, then the gamma is similar to the chi-squared distribution (bottom and upper right panels).

shown that the vector $(Y_1/V, Y_2/V, Y_3/V)$ is Dirichlet with vector of parameters the α's, where V is the sum of the Y's (which is also gamma distributed). In NIMBLE, this suggests using $\theta = 1$ to get a uniform prior:

```
...
# transitions: Dirichlet priors with Gamma formulation
for (s in 1:3){
    # Y1, Y2, Y3 for psi11, psi12, psi13
    lpsi1[s] ~ dgamma(alpha[s], 1)
    psi1[s] <- lpsi1[s]/V1 # psi11, psi12, psi13
```

```
  # Y'1, Y'2, Y'3 for psi21, psi22, psi23
  lpsi2[s] ~ dgamma(alpha[s], 1)
  psi2[s] <- lpsi2[s]/V2 # psi21, psi22, psi23
  # Y''1, Y''2, Y''3 for psi31, psi32, psi33
  lpsi3[s] ~ dgamma(alpha[s], 1)
  psi3[s] <- lpsi3[s]/V3 # psi31, psi32, psi33
}
V1 <- sum(lpsi1[1:3])
V2 <- sum(lpsi2[1:3])
V3 <- sum(lpsi3[1:3])

...
```

From there, we can express the shape parameter of the gamma distribution (precisely the α's here) as a function of covariates as in $\log(\alpha) = \beta_1 + \beta_2 x$ in the spirit of a generalized linear model with a gamma response.

5.4.2 Multinomial logit

Another possibility to build a prior that ensures the movement probabilities are between 0 and 1 and sum up to 1 is to extend the logit link we used for the CJS model in Section 4.8. Remember we had $\text{logit}(\phi) = \beta$, we specified a prior on β say $\beta \sim N(0, 1.5)$, then got a prior on ϕ by back-transforming β with $\phi = \text{logit}^{-1}(\beta)$.

Going back to our example with 3 sites, and focusing on the movement probabilities say, from site 1, we first choose a reference (or pivot) site, say 1, then $\log\left(\dfrac{\psi^{12}}{\psi^{11}}\right) = \beta_2$ and $\log\left(\dfrac{\psi^{13}}{\psi^{11}}\right) = \beta_3$. Interestingly, when exponentiated, the β's here can be interpreted as the increase in the odds of moving versus staying on site resulting from a one-unit increase in the covariate. Any of the sites can be chosen to be the reference, this will not change the likelihood and you will get the same results. Now we specify a normal prior distribution for the β's. Eventually, to back-transform, we use $\psi^{12} = \dfrac{\exp(\beta_2)}{1 + \exp(\beta_2) + \exp(\beta_3)}$ and

$\psi^{13} = \dfrac{\exp(\beta_3)}{1 + \exp(\beta_2) + \exp(\beta_3)}$. The reference parameter, here ψ^{11}, is

calculated as $\psi^{11} = \dfrac{1}{1 + \exp(\beta_2) + \exp(\beta_3)}$, or simply as the comple-

mentary probability $\psi^{11} = 1 - \psi^{12} - \psi^{13}$.

Note that when there are only 2 sites instead of 3 or more, then the multinomial logit reduces to the logit link.

In NIMBLE, we write:

```r
multisite <- nimbleCode({
...
  # transitions: multinomial logit
  for (i in 1:2){
    # normal priors on logit of all but one movement prob
    beta1[i] ~ dnorm(0, sd = 1.5)
    beta2[i] ~ dnorm(0, sd = 1.5)
    beta3[i] ~ dnorm(0, sd = 1.5)
    # constrain the transitions such that their sum is < 1
    psi1[i] <- exp(beta1[i]) / (1 + exp(beta1[1]) + exp(beta1[2]))
    psi2[i] <- exp(beta2[i]) / (1 + exp(beta2[1]) + exp(beta2[2]))
    psi3[i] <- exp(beta3[i]) / (1 + exp(beta3[1]) + exp(beta3[2]))
  }
  # reference movement probability
  psi1[3] <- 1 - psi1[1] - psi1[2]
  psi2[3] <- 1 - psi2[1] - psi2[2]
  psi3[3] <- 1 - psi3[1] - psi3[2]
...
```

Then we use these parameters (which now respect the constraints) to define the transition matrix:

```r
multisite <- nimbleCode({
...
  # probabilities of state z(t+1) given z(t)
  gamma[1,1] <- phi1 * psi1[1]
  gamma[1,2] <- phi1 * psi1[2]
```

```
gamma[1,3] <- phi1 * psi1[3]
gamma[1,4] <- 1 - phi1
gamma[2,1] <- phi2 * psi2[1]
gamma[2,2] <- phi2 * psi2[2]
gamma[2,3] <- phi2 * psi2[3]
gamma[2,4] <- 1 - phi2
gamma[3,1] <- phi3 * psi3[1]
gamma[3,2] <- phi3 * psi3[2]
gamma[3,3] <- phi3 * psi3[3]
gamma[3,4] <- 1 - phi3
gamma[4,1] <- 0
gamma[4,2] <- 0
gamma[4,3] <- 0
gamma[4,4] <- 1

...
```

You may check that the results are very similar to those we obtained
with the Dirichlet prior:

```
MCMCsummary(mcmc.multisite, round = 2)
##             mean    sd 2.5%   50% 97.5% Rhat n.eff
## p1          0.52 0.09 0.36 0.52  0.69 1.02   702
## p2          0.46 0.05 0.37 0.46  0.57 1.00   538
## p3          0.23 0.06 0.13 0.23  0.36 1.00   404
## phi1        0.61 0.05 0.51 0.61  0.71 1.00   893
## phi2        0.70 0.04 0.63 0.70  0.77 1.00   879
## phi3        0.77 0.07 0.64 0.77  0.91 1.01   462
## psi1[1]     0.73 0.06 0.62 0.74  0.83 1.02  1435
## psi1[2]     0.22 0.05 0.13 0.22  0.33 1.00  1687
## psi1[3]     0.05 0.03 0.01 0.04  0.12 1.03   313
## psi2[1]     0.07 0.02 0.04 0.07  0.12 1.00  1318
## psi2[2]     0.83 0.04 0.73 0.84  0.90 1.01   466
## psi2[3]     0.10 0.04 0.04 0.09  0.18 1.00   325
## psi3[1]     0.03 0.02 0.01 0.02  0.07 1.01  2526
## psi3[2]     0.21 0.05 0.12 0.21  0.33 1.00  1197
## psi3[3]     0.76 0.06 0.64 0.76  0.86 1.00   994
```

Both the Dirichlet prior and the multinomial logit link give similar results, and there is not much difference in terms of runtime or quality of convergence, so you should use the option you feel the most comfortable with.

5.5 Sites may be states

So far, we have considered geographical locations (or sites) to refine the alive information when an animal is detected. However, it was quickly realized that sites could actually be states defined by physiology or behavior, hence opening up an avenue for applications of capture-recapture models in many fields of ecology.

Examples of states include:

- Epidemiological or disease states: sick/healthy, uninfected/infected/recovered;

- Morphological states: small/medium/big, light/medium/heavy;

- Life-history states: e.g. breeder/non-breeder, failed breeder, first-time breeder;

- Developmental states: e.g. juvenile/subadult/adult;

- Social states: e.g. solitary/group-living, subordinate/dominant;

- Death states: e.g. alive, dead from harvest, dead from natural causes.

In brief, states are individual, time-specific discrete covariates.

5.5.1 Titis data

To illustrate this section, we will consider data collected between 1942 and 1956 by Lance Richdale on the Sooty shearwaters (*Ardenna grisea*), also known as titis (Figure 5.4).

FIGURE 5.4: Sooty shearwater (*Ardenna grisea*). Credit: John Harrison.

You may see the data below:

```
titis <- read_csv2("titis.csv",
                   col_names = FALSE)
titis %>%
  rename(year_1942 = X1,
         year_1943 = X2,
         year_1944 = X3,
         year_1949 = X4,
         year_1952 = X5,
         year_1953 = X6,
         year_1956 = X7)


## # A tibble: 1,013 x 7
##    year_1942 year_1943 year_1944 year_1949 year_1952
##        <dbl>     <dbl>     <dbl>     <dbl>     <dbl>
## 1          0         0         0         0         0
## 2          0         0         0         0         0
```

```
##  3              0            0            0            0            0
##  4              0            0            0            0            0
##  5              0            0            0            0            0
##  6              0            0            0            0            0
##  7              0            0            0            0            0
##  8              0            0            0            0            0
##  9              0            0            0            0            0
## 10              0            0            0            0            0
## # i 1,003 more rows
## # i 2 more variables: year_1953 <dbl>, year_1956 <dbl>
```

In total, 1013 titis were captured, marked and recaptured on a small
colony on Whero Island in southern New Zealand. These data were
previously analyzed by Richard Scofield who kindly provided us with
the data.

Following the way the data were collected, four states were originally
considered: Alive breeder; Accompanied by another bird in a burrow;
Alone in a burrow; On the surface; Dead. For simplicity, we pooled
all alive states (except breeder) together in a non-breeder state (NB)
that includes failed breeders (birds that had bred previously – skip
reproduction or divorce) and pre-breeders (birds that had yet to breed).
Because burrows were not checked before hatching, some birds in the
category NB might have already failed. Therefore, birds in the breeder
state (B) should be seen as successful breeders, and those in the NB
state as nonbreeders plus prebreeders and failed breeders.

In summary, we code the states as 1 for alive and breeding, 2 for alive
and non-breeding and 3 for dead. To make the modeling process more
self-explanatory, we will use letters B, NB and D but you should keep in
mind that these are actually numbers. Observations are non-detected
coded as 1, and detected as breeder or non-breeder coded as 2 and 3
respectively.

The aim here is to study life-history trade–offs. Specifically, we ask the
questions: Does breeding affect survival? Does breeding in current year
affect breeding next year?

5.5.2   The AS model for states

Basically, the AS model with states is the same model as in the geese example with two sites, see Sections 5.2 and 5.3.

If we let ϕ^B and ϕ^{NB} be the survival probabilities of breeders and non-breeders, ψ^{NBB} the probability of becoming breeder for a non-breeder, and ψ^{BNB} the probability of skipping reproduction, then the transition matrix is:

$$\Gamma = \begin{pmatrix} \phi^B(1-\psi^{BNB}) & \phi^B\psi^{BNB} & 1-\phi^B \\ \phi^{NB}\psi^{NBB} & \phi^{NB}(1-\psi^{NBB}) & 1-\phi^{NB} \\ 0 & 0 & 1 \end{pmatrix} \begin{matrix} z_{t-1}=B \\ z_{t-1}=NB \\ z_{t-1}=D \end{matrix}$$

$$\begin{matrix} z_t=B & z_t=NB & z_t=D \end{matrix}$$

The costs or reproduction would reflect in future reproduction if breeders have a lower probability of breed next year than non-breeders $\psi^{BB} = 1 - \psi^{BNB} < \psi^{NBB}$ or in survival if the survival of breeders is lower than that of non-breeders $\phi^B < \phi^{NB}$.

The observation matrix is:

$$\Omega = \begin{pmatrix} 1-p^B & p^B & 0 \\ 1-p^{NB} & 0 & p^{NB} \\ 1 & 0 & 0 \end{pmatrix} \begin{matrix} z_t=B \\ z_t=NB \\ z_t=D \end{matrix}$$

$$\begin{matrix} y_t=1 & y_t=2 & y_t=3 \end{matrix}$$

where p^B and p^{NB} are the detection probabilities of non-breeders and breeders respectively.

5.5.3   NIMBLE implementation

We first write the NIMBLE code, which is exactly the same as in the geese example with two sites, see Section 5.3.

First some definitions which we have as comments in the code:

```r
multistate <- nimbleCode({
  # ---------------------------------------------------------
  # Parameters:
  # B is for breeder, NB for non-breeder
  # phiB: survival probability state B
  # phiNB: survival probability state NB
  # psiBNB: transition probability from B to NB
  # psiNBB: transition probability from NB to B
  # pB: detection probability B
  # pNB: detection probability NB
  # ---------------------------------------------------------
  # States (z):
  # 1 alive B
  # 2 alive NB
  # 3 dead
  # Observations (y):
  # 1 not seen
  # 2 seen as B
  # 3 seen as NB
  # ---------------------------------------------------------
...
```

Then the priors:

```r
multistate <- nimbleCode({
...
  # Priors
  phiB ~ dunif(0, 1)
  phiNB ~ dunif(0, 1)
  psiBNB ~ dunif(0, 1)
  psiNBB ~ dunif(0, 1)
  pB ~ dunif(0, 1)
  pNB ~ dunif(0, 1)
...
```

The transition matrix:

```r
multistate <- nimbleCode({

...

  # probabilities of state z(t+1) given z(t)
  gamma[1,1] <- phiB * (1 - psiBNB)
  gamma[1,2] <- phiB * psiBNB
  gamma[1,3] <- 1 - phiB
  gamma[2,1] <- phiNB * psiNBB
  gamma[2,2] <- phiNB * (1 - psiNBB)
  gamma[2,3] <- 1 - phiNB
  gamma[3,1] <- 0
  gamma[3,2] <- 0
  gamma[3,3] <- 1

...
```

The observation matrix:

```r
multistate <- nimbleCode({

...

  # probabilities of y(t) given z(t)
  omega[1,1] <- 1 - pB      # Pr(alive B t -> non-detected t)
  omega[1,2] <- pB          # Pr(alive B t -> detected B t)
  omega[1,3] <- 0           # Pr(alive B t -> detected NB t)
  omega[2,1] <- 1 - pNB     # Pr(alive NB t -> non-detected t)
  omega[2,2] <- 0           # Pr(alive NB t -> detected B t)
  omega[2,3] <- pNB         # Pr(alive NB t -> detected NB t)
  omega[3,1] <- 1           # Pr(dead t -> non-detected t)
  omega[3,2] <- 0           # Pr(dead t -> detected N t)
  omega[3,3] <- 0           # Pr(dead t -> detected NB t)

...
```

And the likelihood:

```r
multistate <- nimbleCode({

...

  # likelihood
```

```r
  for (i in 1:N){
    # latent state at first capture
    z[i,first[i]] <- y[i,first[i]] - 1
    for (t in (first[i]+1):K){
      # z(t) given z(t-1)
      z[i,t] ~ dcat(gamma[z[i,t-1],1:3])
      # y(t) given z(t)
      y[i,t] ~ dcat(omega[z[i,t],1:3])
    }
  }
})
```

We run NIMBLE and get the following results:

```r
MCMCsummary(mcmc.multistate, round = 2)
##           mean   sd 2.5%  50% 97.5% Rhat n.eff
## pB        0.60 0.03 0.54 0.60  0.65 1.00   756
## pNB       0.56 0.03 0.51 0.56  0.62 1.01   725
## phiB      0.80 0.02 0.77 0.80  0.83 1.00  1485
## phiNB     0.85 0.02 0.81 0.85  0.88 1.01  1231
## psiBNB    0.25 0.02 0.21 0.25  0.30 1.01  1227
## psiNBB    0.24 0.02 0.20 0.24  0.28 1.00  1123
```

Non-breeder individuals seem to have a survival higher than breeder individuals, suggesting a trade-off between reproduction and survival. Let's compare graphically the survival of breeder and non-breeder individuals. First we gather the values generated for ϕ^B and ϕ^{NB} for the two chains:

```r
phiB <- c(mcmc.multistate$chain1[,"phiB"],
          mcmc.multistate$chain2[,"phiB"])
phiNB <- c(mcmc.multistate$chain1[,"phiNB"],
           mcmc.multistate$chain2[,"phiNB"])
df <- data.frame(param = c(rep("phiB", length(phiB)),
```

```
                               rep("phiNB", length(phiB))),
                   value = c(phiB, phiNB))
```

Then, we plot the two posterior distributions:

```
df %>%
  ggplot(aes(x = value, fill = param)) +
  geom_density(color = "white", alpha = 0.6, position = 'identity') +
  scale_fill_manual(values = c("#69b3a2", "#404080")) +
  labs(fill = "", x = "survival")
```

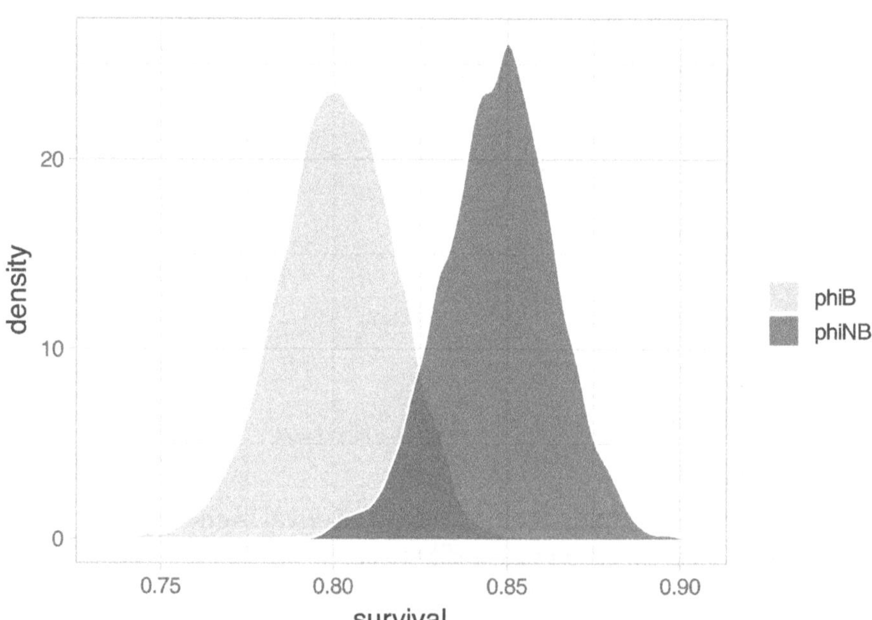

There is little overlap between the two distributions, suggesting an
actual trade–off. A formal test of the trade-off would consist in fitting
the model with survival irrespective of the state, and compare its WAIC
value to the model we just fitted.

What about a potential trade-off on reproduction?

```
psiBNB <- c(mcmc.multistate$chain1[,"psiBNB"],
            mcmc.multistate$chain2[,"psiBNB"])
```

```r
psiBB <- 1 - psiBNB
psiNBB <- c(mcmc.multistate$chain1[,"psiNBB"],
            mcmc.multistate$chain2[,"psiNBB"])
df <- data.frame(param = c(rep("psiBB", length(phiB)),
                           rep("psiNBB", length(phiB))),
                 value = c(psiBB, psiNBB))
df %>%
  ggplot(aes(x = value, fill = param)) +
  geom_density(color = "white", alpha = 0.6, position = 'identity') +
  scale_fill_manual(values = c("#69b3a2", "#404080")) +
  labs(fill = "", x = "breeding probabilities")
```


There is no overlap whatsoever, so the two transition probabilities are clearly different. Interestingly, breeder individuals do much better than non-breeder individuals. This failure at detecting a trade-off is probably due to individual heterogeneity that should be accounted for. You could add an individual random effect as in Section 4.8.4 or consider 2 classes of individuals as we will do in a case study at Section 7.4.

5.6   Issue of local minima

In the frequentist approach, we use the maximum likelihood theory to estimate parameters. The maximum likelihood estimates are the values that get you to the maximum of the model likelihood. To find out the maximum of the likelihood, we use iterative optimization algorithms (e.g. the default method is that of Nelder and Mead in the R `optim()` function). However, sometimes, our model likelihood contains several maxima and there is no guarantee that the algorithms will find the global maximum corresponding to the maximum likelihood estimates, and it may get stuck in a local maximum. Let's illustrate this issue with some simulated data that were kindly provided by Jérôme Dupuis. We consider 2 sites (or alive states), say 1 and 2, and 7 sampling occasions. The survival probability is constant $\phi = 1$ as well as the detection probability $p = 0.6$. The probability of moving from 1 to 2 is $\psi^{12} = 0.6$ and $\psi^{21} = 0.85$ in the opposite direction. Here are the encounter histories of the 27 individuals that were simulated by Jérôme:

```r
dat <- matrix(c(2, 0, 2, 1, 2, 0, 2,
                2, 0, 2, 1, 2, 0, 2,
                2, 0, 2, 1, 2, 0, 2,
                2, 0, 2, 1, 2, 0, 2,
                1, 1, 1, 0, 1, 0, 1,
                1, 1, 1, 0, 1, 0, 1,
                1, 1, 1, 0, 1, 0, 1,
                1, 1, 1, 0, 1, 0, 1,
                2, 0, 2, 0, 2, 0, 1,
                2, 0, 2, 0, 2, 0, 1,
                2, 0, 2, 0, 2, 0, 1,
                2, 0, 2, 0, 2, 0, 1,
                1, 0, 1, 0, 1, 0, 1,
                1, 0, 1, 0, 1, 0, 1,
                1, 0, 1, 0, 1, 0, 1,
                1, 0, 1, 0, 1, 0, 1,
                2, 0, 2, 0, 2, 0, 2,
                2, 0, 2, 0, 2, 0, 2,
                2, 0, 2, 0, 2, 0, 2,
```

```
                2, 0, 2, 0, 2, 0, 2,
                1, 0, 1, 0, 1, 0, 2,
                1, 0, 1, 0, 1, 0, 2,
                1, 0, 1, 0, 1, 0, 2,
                1, 0, 1, 0, 1, 0, 2,
                2, 2, 0, 1, 0, 2, 1,
                2, 2, 0, 1, 0, 2, 1,
                2, 2, 0, 1, 0, 2, 1,
                2, 2, 0, 1, 0, 2, 1,
                2, 1, 0, 2, 0, 1, 1,
                2, 1, 0, 2, 0, 1, 1,
                2, 1, 0, 2, 0, 1, 1,
                2, 1, 0, 2, 0, 1, 1),
            byrow = T,
            ncol = 7)
```

In Figure 5.5, we provide an illustration of the influence of the choice
of initial values when trying to maximize the likelihood, or rather to
minimize the deviance (which is minus two times the log of the likelihood). The black curve is the what we called the profile deviance for ψ^{21}.
Profiling the deviance consists in taking a slice of it in the direction of
a parameter of interest and treating the other parameters as nuisance
parameters. In our example, we set ψ^{21} to a value (on the x-axis) an minimize the deviance (on the y-axis) with respect to the other parameters.
There are two minima, but only the global minimum (corresponding
to the lowest value of deviance) corresponding to ψ^{21} around 0.8 is
of interest to us. The thing is that if you start your optimization algorithm by picking value in the red area, then it will get stuck in the local
minimum and will tell you the maximum likelihood estimate of ψ^{21} is
around 0.35, which is obviously far from the value we used to simulate
the data. In contrast, if you pick initial values in the green area, then
the algorithm will converge to the global minimum.

You might argue that this is a problem of the optimization algorithm
and therefore inherent to the frequentist approach. Well, it turns out
that MCMC algorithms are not immune to the issue. If you fit the AS
model with constant parameters to the simulated data, here is the trace
for the probability of moving from 2 to 1:

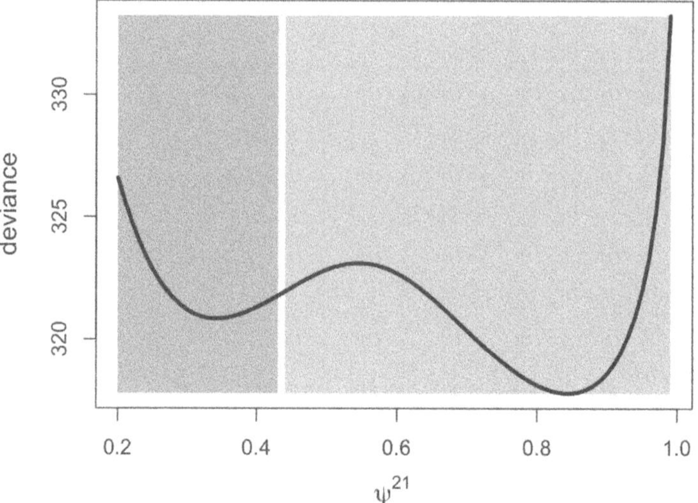

FIGURE 5.5: Influence of the choice of initial values on the convergence to the global minimum of the deviance illustrated with simulated data. The black curve is the profile deviance of the probability to move from site 2 to 1. If an initial value is picked in the left area, we end up in the local minimum while if it is picked in the green area, then we get the global minimum which corresponds to the maximum likelihood estimate.

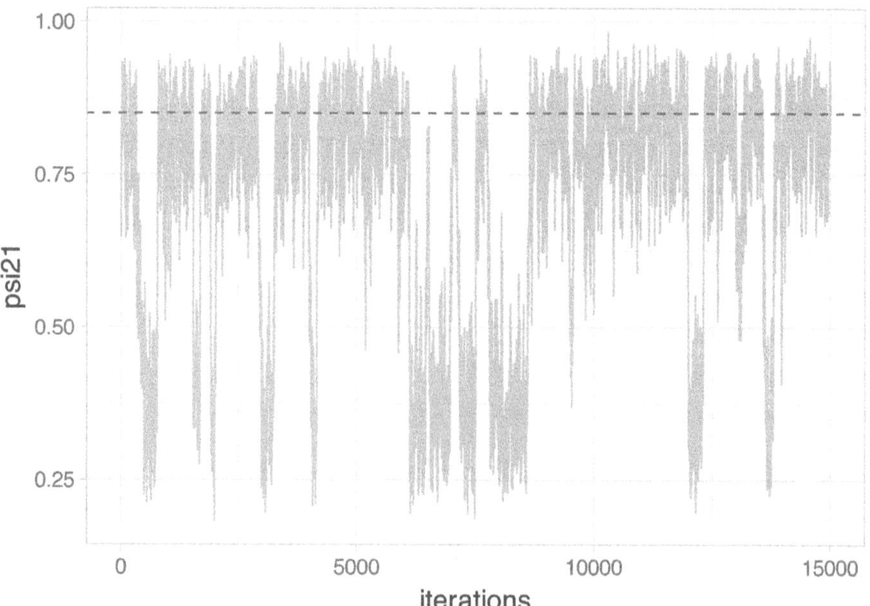

Clearly, there are two regimes. The chain spends most of its time around high values of ψ^{21} close to the true value represented by the dashed line. But sometimes, the chain jumps to values around 0.3–0.4. This behavior translates into two modes in the posterior distribution for ψ^{21} where the mode on the right is closer to the truth represented by the dashed vertical line:

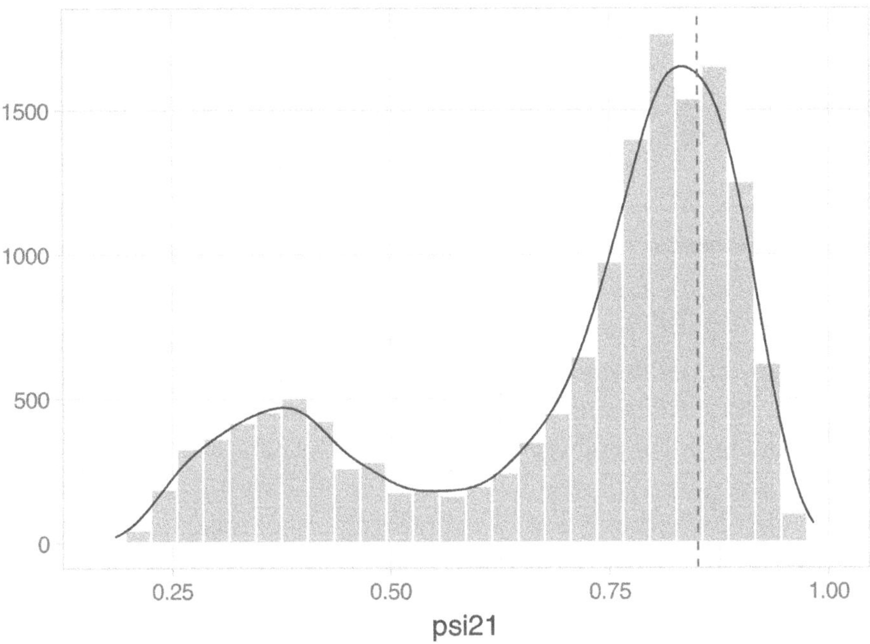

The issue of local minima is a difficult problem. How to get out of this problematic situation? In the frequentist approach, the trick is to fit your model several times with different initial values each time, hoping that you'll get to fall in the green area somehow as in Figure 5.5. In the Bayesian approach, the key to handle distributions with multiple modes is to sample the posteriors efficiently. Assuming the chains we run in NIMBLE spend more time in the region of the parameter space corresponding to the global minimum, then I recommend using the median or the mode to summarize the posterior distribution. In the simulated example, we get a median of 0.79 for ψ^{21}, not too bad given that the data were simulated with a value of 0.85 for that parameter:

```
MCMCsummary(mcmc.multisite, round = 2)
##          mean    sd 2.5%  50% 97.5% Rhat n.eff
## p        0.58 0.04 0.51 0.58  0.65 1.00  3194
## phi      0.99 0.01 0.98 1.00  1.00 1.02  1173
## psi12    0.51 0.15 0.19 0.56  0.72 1.10    76
## psi21    0.70 0.20 0.27 0.78  0.93 1.12    56
```

As a general advice, I recommend to always inspect the trace plots to find out whether you have posterior distributions with multiple modes which would suggest local minima.

5.7 Uncertainty

In the AS model, we assume that we can without a doubt assign a site or a state to an animal whenever it is detected. But this is not always the case. For example, when the breeding status in mammals or birds is ascertained based on the presence of offspring or eggs, we are uncertain of whether a female is breeding or not when the offspring or eggs are not seen. Another example is when the epidemiological status in mammals or birds is ascertained based on some tests run on some animals when captured, we are uncertain whether these animals are healthy or sick when detected from distance without possibility to manipulate them for testing. In this section, we will cover the extension of the AS model to uncertain states – known as multievent models after Pradel [2005] – through two examples, one on breeding states and the other on disease states. We will cover other examples in the part on Case studies.

5.7.1 Breeding states

We will revisit the titis example 5.5 and try to assess life-history trade-offs while accounting for uncertainty in breeding status. We still have 3 states, which are alive and breeding, alive and non-breeding and dead. With regard to observations, a bird may be not encountered. It may also be encountered, but in contrast with our previous analysis of the titis data, we don't know its state for sure. It may be found and ascertained (or classified) as breeder. It may be found and ascertained as non-breeder. It may be found but we are unable to determine whether it's breeding or non-breeding.

How do the states generate the observations?

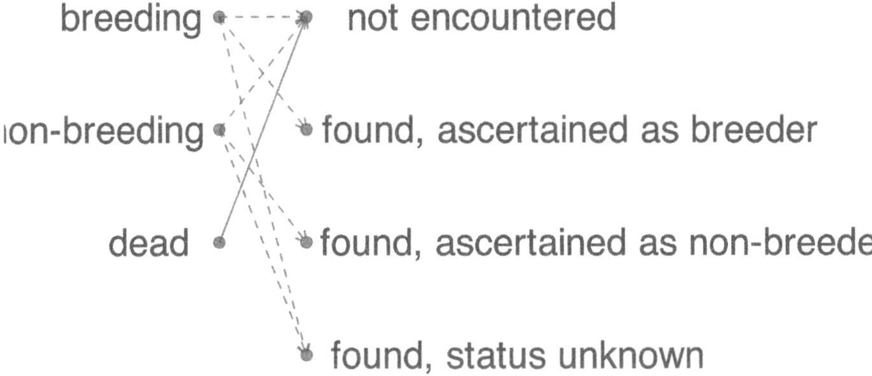

Each alive state can generate 3 observations. The only deterministic link is that between the dead state and the observation non-encountered, because if a bird is dead, it cannot be detected for sure.

Let's specify the model. First thing we need, and it's a big difference with the AS model, we need initial state probabilities because we cannot assign states to individuals with certainty. We write down the probability for each state at first encounter, or the vector of initial state probabilities:

$$\delta = \begin{pmatrix} z_t = B & z_t = NB & z_t = D \\ \pi^B & 1 - \pi^B & 0 \end{pmatrix}$$

where π^B is the probability that a newly encountered individual is a breeder, and $\pi^{NB} = 1 - \pi^B$ is the probability that a newly encountered individual is a non-breeder (the complementary probability of π^B). The probability of being dead at first encounter is 0 (a bird is alive when it is first encountered).

Now the transition matrix, this is the easy part as it doesn't change. We have:

$$
\Gamma = \begin{array}{cccc}
 & z_t = B & z_t = NB & z_t = D \\
\begin{pmatrix} \phi^B(1 - \psi^{BNB}) & \phi^B\psi^{BNB} & 1 - \phi^B \\ \phi^{NB}\psi^{NBB} & \phi^{NB}(1 - \psi^{NBB}) & 1 - \phi^{NB} \\ 0 & 0 & 1 \end{pmatrix} & \begin{array}{c} z_{t-1} = B \\ z_{t-1} = NB \\ z_{t-1} = D \end{array}
\end{array}
$$

where ϕ^B is the breeder survival, ϕ_{NB} that of non-breeders, ψ^{BNB} is the probability for an individual breeding a year to be a non-breeder the next year, and ψ^{NBB} is the probability for an non-breeder individual to breeder the next year.

Last, the observation matrix. The main difference between multisite/multistate and multievent models is here, in the observation parameters. Besides p^B the detection probability of breeders and p^{NB} that of non-breeders, we introduce two new parameters: β^B is the probability to correctly assign an individual that is in state B to state B, and β^{NB} is the probability to correctly assign an individual that is in state NB to state NB. The complementary of these β parameters are often called false positive probabilities. We put everything in a matrix, as usual. In rows we have the states: breeding, non-breeding and dead. In columns, at the same occasion, we have the observations: non-detected, detected and ascertained B, detected and ascertained NB, and detected but state unknown:

$$
\Omega = \begin{array}{ccccc}
 & y_t = 1 & y_t = 2 & y_t = 3 & y_t = 4 \\
\begin{pmatrix} 1 - p^B & p^B\beta^B & 0 & p^B(1 - \beta_B) \\ 1 - p^{NB} & 0 & p^{NB}\beta^{NB} & p^{NB}(1 - \beta^{NB}) \\ 1 & 0 & 0 & 0 \end{pmatrix} & \begin{array}{c} z_t = B \\ z_t = NB \\ z_t = D \end{array}
\end{array}
$$

For example, the probability of being detected and assigned to state B, given that you're in state B is the product of p^B the detection probability in B and β^B the probability of correctly assigning a breeding individual to state B.

At first encounter, all individuals are captured, but you still need to assign them a state. This means that we should set $p^B = p^{NB} = 1$ and use:

$$
\begin{array}{cccc}
y_{t=\text{first}}=1 & y_{t=\text{first}}=2 & y_{t=\text{first}}=3 & y_{t=\text{first}}=4 \\
\end{array}
$$

$$
\left(
\begin{array}{cccc}
0 & \beta^B & 0 & (1-\beta^B) \\
0 & 0 & \beta^{NB} & (1-\beta^{NB}) \\
1 & 0 & 0 & 0
\end{array}
\right)
\begin{array}{l}
z_{t=\text{first}}=B \\
z_{t=\text{first}}=NB \\
z_{t=\text{first}}=D
\end{array}
$$

To implement this model in NIMBLE, we start by using comments to define the parameters, states and observations in the header of the code:

```
multievent <- nimbleCode({
  # --------------------------------------------------------------
  # Parameters:
  # phiB: survival probability state B
  # phiNB: survival probability state NB
  # psiBNB: transition probability from B to NB
  # psiNBB: transition probability from NB to B
  # pB: recapture probability B
  # pNB: recapture probability NB
  # piB prob. of being in initial state breeder
  # betaNB prob ascertain breeding status of ind encountered as NB
  # betaB prob ascertain breeding status of ind encountered as B
  # --------------------------------------------------------------
  # States (z):
  # 1 alive B
  # 2 alive NB
  # 3 dead
  # Observations (y):
  # 1 = non-detected
  # 2 = seen and ascertained as breeder
  # 3 = seen and ascertained as non-breeder
  # 4 = not ascertained
  # --------------------------------------------------------------
...
```

Then we assign prior to all parameters to be estimated. Because we deal with probabilities, the uniform distribution between 0 and 1 will do the job:

```
multievent <- nimbleCode({
...
  # Priors
  phiB ~ dunif(0, 1)
  phiNB ~ dunif(0, 1)
  psiBNB ~ dunif(0, 1)
  psiNBB ~ dunif(0, 1)
  pB ~ dunif(0, 1)
  pNB ~ dunif(0, 1)
  piB ~ dunif(0, 1)
  betaNB ~ dunif(0, 1)
  betaB ~ dunif(0, 1)
...
```

Now we write the vector of initial state probabilities:

```
multievent <- nimbleCode({
...
  # vector of initial stats probs
  delta[1] <- piB # prob. of being in initial state B
  delta[2] <- 1 - piB # prob. of being in initial state NB
  delta[3] <- 0 # prob. of being in initial state dead
...
```

The transition matrix:

```
multievent <- nimbleCode({
...
  # probabilities of state z(t+1) given z(t)
  gamma[1,1] <- phiB * (1 - psiBNB)
  gamma[1,2] <- phiB * psiBNB
  gamma[1,3] <- 1 - phiB
  gamma[2,1] <- phiNB * psiNBB
```

```
  gamma[2,2] <- phiNB * (1 - psiNBB)
  gamma[2,3] <- 1 - phiNB
  gamma[3,1] <- 0
  gamma[3,2] <- 0
  gamma[3,3] <- 1
...
```

And the observation matrix:

```
multievent <- nimbleCode({
...
  # probabilities of y(t) given z(t)
  omega[1,1] <- 1 - pB              # Pr(alive B t -> non-detected t)
  omega[1,2] <- pB * betaB          # Pr(alive B t -> detected B t)
  omega[1,3] <- 0                   # Pr(alive B t -> detected NB t)
  omega[1,4] <- pB * (1 - betaB)    # Pr(alive B t -> detected U t)
  omega[2,1] <- 1 - pNB             # Pr(alive NB t -> non-detected t)
  omega[2,2] <- 0                   # Pr(alive NB t -> detected B t)
  omega[2,3] <- pNB * betaNB        # Pr(alive NB t -> detected NB t)
  omega[2,4] <- pNB * (1 - betaNB)  # Pr(alive NB t -> detected U t)
  omega[3,1] <- 1                   # Pr(dead t -> non-detected t)
  omega[3,2] <- 0                   # Pr(dead t -> detected N t)
  omega[3,3] <- 0                   # Pr(dead t -> detected NB t)
  omega[3,4] <- 0                   # Pr(dead t -> detected U t)
...
```

The observation matrix at first encounter:

```
multievent <- nimbleCode({
...
  # probabilities of y(first) given z(first)
  # Pr(alive B t = first -> non-detected t = first)
  omega.init[1,1] <- 0
  # Pr(alive B t = first -> detected B t = first)
  omega.init[1,2] <- betaB
  # Pr(alive B t = first -> detected NB t = first)
  omega.init[1,3] <- 0
  # Pr(alive B t = first -> detected U t = first)
```

```r
  omega.init[1,4] <- 1 - betaB
  # Pr(alive NB t = first -> non-detected t = first)
  omega.init[2,1] <- 0
  # Pr(alive NB t = first -> detected B t = first)
  omega.init[2,2] <- 0
  # Pr(alive NB t = first -> detected NB t = first)
  omega.init[2,3] <- betaNB
  # Pr(alive NB t = first -> detected U t = first)
  omega.init[2,4] <- 1 - betaNB
  # Pr(dead t = first -> non-detected t = first)
  omega.init[3,1] <- 1
  # Pr(dead t = first -> detected N t = first)
  omega.init[3,2] <- 0
  # Pr(dead t = first -> detected NB t = first)
  omega.init[3,3] <- 0
  # Pr(dead t = first -> detected U t = first)
  omega.init[3,4] <- 0
...
```

Eventually, we get to the likelihood:

```r
multievent <- nimbleCode({
...
  # likelihood
  for (i in 1:N){
    # latent state at first capture
    z[i,first[i]] ~ dcat(delta[1:3])
    # obs at first encounter
    y[i,first[i]] ~ dcat(omega.init[z[i,first[i]],1:4])
    for (t in (first[i]+1):K){
      # z(t) given z(t-1)
      z[i,t] ~ dcat(gamma[z[i,t-1],1:3])
      # y(t) given z(t)
      y[i,t] ~ dcat(omega[z[i,t],1:4])
    }
  }
})
```

The only change is in the line `y[i,first[i]] ~ dcat(omega.init[z[i,first[i]],1:4])` where we use the observation matrix at first encounter.

We run NIMBLE and get the following numerical summaries for the model parameters:

```
MCMCsummary(mcmc.multievent, round = 2)
##            mean    sd  2.5%   50%  97.5%  Rhat  n.eff
## betaB      0.19  0.01  0.16  0.19   0.22  1.00    583
## betaNB     0.74  0.06  0.64  0.73   0.88  1.00    198
## pB         0.56  0.03  0.51  0.56   0.62  1.01    801
## pNB        0.60  0.04  0.53  0.60   0.67  1.02    757
## phiB       0.81  0.02  0.78  0.81   0.85  1.00   1551
## phiNB      0.84  0.02  0.80  0.84   0.87  1.00   1610
## piB        0.70  0.03  0.64  0.70   0.76  1.00    296
## psiBNB     0.22  0.03  0.17  0.22   0.28  1.01    925
## psiNBB     0.23  0.05  0.15  0.23   0.35  1.00    250
```

Breeders are difficult to assign to the correct state β^B, while non-breeders are relatively well classified as non-breeders β^{NB}.

There is again no cost of current reproduction on future reproduction. We no longer detect a cost of breeding on survival.

5.7.2 Disease states

Let's have a look to another example. We consider a system of an emerging pathogen *Mycoplasma gallisepticum* Edward and Kanarek and its host the house finch, *Carpodacus mexicanus* Müller. The pathogen causes moderate to severe eye swelling (Figure 5.6).

Here we're asking whether the presence of clinical signs of the pathogen influences survival. The objective of the study was also to quantify infection (moving from state healthy to state ill) and recovery (moving from state ill to state healthy) probabilities. The birds were captured via mist nets and marked with individually identifiable color bands over three years. The data were kindly provided by Paul Conn and Evan Cooch. The difficulty was that ascertaining the disease status of birds

FIGURE 5.6: A house finch with a heavy infection caused by conjunc-
tivitis. Credit: Jim Mondok.

seen from distance was difficult since determining the presence of the
pathogen was only possible when the bird's eyes were clearly visible. In
this context, how to study the dynamics of the disease?

First, we think of states and observations. We have:

- 3 states
    - healthy (H)
    - ill (I)
    - dead (D)
- 4 observations
    - not seen (1)
    - captured healthy (2)
    - captured ill (3)
    - health status unknown, i.e. seen at distance (4)

How do the states generate observations?

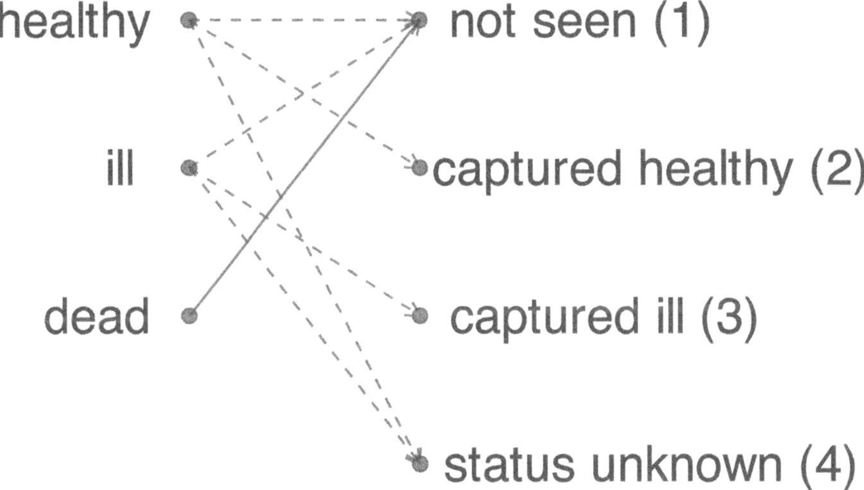

Clearly, this is the same model as in the previous section on this, Section 5.7.1, in which loosely speaking we replace breeder by healthy and non-breeder by ill.

The vector of initial state probabilities is:

$$\delta = \begin{pmatrix} z_t = H & z_t = I & z_t = D \\ \pi^H & 1 - \pi^H & 0 \end{pmatrix}$$

where π^H is the probability that a newly encountered individual is healthy, and $\pi^I = 1 - \pi^H$ is the probability that a newly encountered individual is ill.

The transition matrix is:

$$\Gamma = \begin{pmatrix} z_t = H & z_t = I & z_t = D & \\ \phi^H(1 - \psi^{HI}) & \phi^H \psi^{HI} & 1 - \phi^H & z_{t-1} = H \\ \phi^I \psi^{IH} & \phi^I(1 - \psi^{IH}) & 1 - \phi^I & z_{t-1} = I \\ 0 & 0 & 1 & z_{t-1} = D \end{pmatrix}$$

where ϕ^H is the survival probability of healthy individuals, ϕ^I the survival probability of ill individuals, ψ^{HI} the probability of getting ill (infection rate) and ψ^{IH} the probability of recovering from the disease (recovery rate).

Image you'd like to model the dynamic of an incurable disease, the transition matrix would be modified by having $\psi^{IH} = 0$, and once a bird gets ill, it remains ill $\psi^{II} = 1 - \psi^{IH} = 1$. Therefore, we would have:

$$\Gamma = \begin{pmatrix} \phi^H(1 - \psi^{HI}) & \phi^H\psi^{HI} & 1 - \phi^H \\ 0 & \phi^I & 1 - \phi^I \\ 0 & 0 & 1 \end{pmatrix} \begin{matrix} z_{t-1} = H \\ z_{t-1} = I \\ z_{t-1} = D \end{matrix}$$

$$\begin{matrix} z_t = H & z_t = I & z_t = D \end{matrix}$$

For analyzing the house finch data, we allow recovering from the disease. The observation matrix is:

$$\begin{matrix} y_t = 0 & y_t = 1 & y_t = 2 & y_t = 3 \end{matrix}$$

$$\Omega = \begin{pmatrix} 1 - p^H & p^H\beta^H & 0 & p^H(1 - \beta^H) \\ 1 - p^I & 0 & p^I\beta^I & p^I(1 - \beta^I) \\ 1 & 0 & 0 & 0 \end{pmatrix} \begin{matrix} z_t = H \\ z_t = I \\ z_t = D \end{matrix}$$

where β^H is the probability to assign a healthy individual to state H, and β^I is the probability to assign a sick individual to state I. p^H is the detection probability of healthy individuals, p^I that of sick individuals.

Using the code we developped for the titis example, we get the following results on the finches by running NIMBLE:

```
##           mean    sd 2.5%  50% 97.5% Rhat n.eff
## betaH 0.99 0.01 0.97 0.99  1.00 1.01  2118
## betaI 0.05 0.01 0.02 0.05  0.08 1.01  1810
## pH    0.71 0.21 0.15 0.75  0.98 1.04   142
## pI    0.28 0.10 0.21 0.25  0.63 1.00   317
## phiH  0.70 0.07 0.62 0.69  0.89 1.08   192
## phiI  0.99 0.01 0.96 0.99  1.00 1.00   832
## pi    0.96 0.01 0.93 0.96  0.98 1.00  6332
## psiHI 0.75 0.17 0.19 0.80  0.88 1.02   160
## psiIH 0.20 0.09 0.12 0.17  0.50 1.05   146
```

Healthy individuals are correctly assigned (β^H is almost 1), while infected individuals are difficult to ascertain (β^I is around 0.05). Unexpectedly, ill birds have a better survival than healthy individuals (compare ϕ^I and ϕ^H). Infection rate (ψ^{HI}) is 75%, recovery rate is 20% (ψ^{IH}).

5.8 Summary

- The AS model is a HMM that extends the CJS model by allowing the estimation of movements between sites (e.g. geographical locations) or transitions between states (e.g. breeding status). This flexibility allows addressing all sorts of questions in ecology and evolution.

- Covariates can be considered, and appropriate priors or link functions need to be used when there are more than 2 sites or 2 alive states.

- Model comparison can be achieved with the WAIC and the goodness of fit of the AS model to capture-recapture data can be assessed with classical procedures.

- Importantly, the HMM framework allows to account for uncertainty when assigning states to individuals.

- The models covered in this chapter haven been called multistratum/strata models, multisite models (Section 5.2), multistate models (Section 5.5) and multievent models (Section 5.7). These models are all HMMs.

5.9 Suggested reading

- The AS model was introduced in Arnason [1972], Arnason [1973] and Schwarz et al. [1993]. Very soon clever folks realized that sites could be replaced by states as in Nichols et al. [1992] and Nichols et al. [1994]. For a review of models with sites and states, see Lebreton et al. [2009].

- The geese data were analyzed in Hestbeck et al. [1991] and Brownie et al. [1993], and the titis data in Scofield et al. [2001]. The house finches data were analyzed in Faustino et al. [2004] and Conn and Cooch [2009; see also Cooch et al., 2012]. Check out Santoro et al. [2014], Marescot et al. [2018] and Ollivier et al. [2023] for other examples in disease ecology.

- Section 5.6 on local minima was inspired by chapter 10 of Cooch and White [2017].

- Classical goodness of fit tests are reviewed in Pradel et al. [2005]. See also Pradel et al. [2003] for tests specifically designed for multi-site/multistate models.

- Models with uncertainty were introduced in Pradel [2005]. Dupuis [1995] had a similar idea for the AS model. For a review, see Gimenez et al. [2012].

Part III

Case studies

Introduction

This third part `Case studies` provides real-world case studies from the scientific literature that you can reproduce using material covered in previous chapters. These problems can either i) be used to cement and deepen your understanding of methods and models, ii) be adapted for your own purpose, or iii) serve as teaching projects. For each case study, I recall the ecological question, build the model step by step and conclude with a discussion of the results. I provide only the code snippets needed to illustrate specific points; the complete code and data are available at https://github.com/oliviergimenez/banana-book/tree/master/appendix.

6

Dealing with covariates

6.1 Introduction

Capture-recapture models often aim not just to estimate demographic parameters, but also to understand what drives their variation as we have seen in Chapter 4. Here, we will cover the selection of covariates from a potentially large set, using an extension of standard MCMC algorithms known as reversible jump MCMC (RJMCMC), and how to handle with uncertainty in key covariates, namely age or sex, when these are not perfectly known for all individuals.

6.2 Covariate selection with Reversible Jump MCMC

6.2.1 Motivation

In Section 4.6, we used WAIC to figure out which model, among several candidates, was best supported by the data. Each model there represented a different ecological hypothesis. Now when you are in a situation where you have several individual or temporal covariates that might explain variation in some demographic parameters, say survival as in Section 4.8, things can quickly become a bit more tedious. Let's look at a real-world example with the white stork (*Ciconia ciconia*) population in Baden-Württemberg, Germany. From the 1960s to the 1990s, all Western European stork populations were in decline. One leading hypothesis was that these declines were driven by reduced food availability, itself caused by severe droughts in the storks' wintering grounds in the Sahel region [e.g. Kanyamibwa et al., 1990, Grosbois et al. [2008]].

To explore this idea, we'll use rainfall measurements from 10 meteorological stations chosen to represent the storks' wintering area: Diourbel, Gao, Kayes, Kita, Maradi, Mopti, Ouahigouya, Ségou, Tahoua, and Tombouctou. The data we have are the cumulative rainfall amounts (in mm) over June, July, August, and September – roughly the rainy season in the Sahel:

```
##         Diourbel  Gao Kayes   Kita Maradi Mopti Ouahigouya
## [1,]        6770 1900  7130  11350   6970  6580       6220
## [2,]        6420 2690  6180  13690   5880  6470       6260
## [3,]        6890 3530  6480  11820   6100  5230       6730
## [4,]        5700 2710  8440   9590   5210  5670       5670
## [5,]        6760 2090  6910   9850   6020  4200       5630
## [6,]        5432 2100  6300   8960   6890  3900       7180
##         Ségou Tahoua Tombouctou
## [1,]     8380   3920       1390
## [2,]     7530   3200       2210
## [3,]     6340   5170       1760
## [4,]     7490   4370       2580
## [5,]     5920   3120       2340
## [6,]     6461   3891       2020
```

Our question is: Which combination of these stations best explains variation in survival? If we tried to answer this by testing every possible model, we would have to consider every subset of the 10 stations and compare 1024 models (2^10 combinations). That's far too many for a WAIC-based approach to be practical.

Instead of comparing all possible models one by one, we can use Reversible Jump MCMC (RJMCMC). RJMCMC is an extension of standard Bayesian inference in which the model itself is treated as an additional, discrete parameter. In this framework, the posterior distribution spans both the parameter space (e.g. regression coefficients) and the model space (e.g. which covariates are included). In simple terms, we allow the algorithm to jump between models, switching covariates on or off as it goes. You can think of it as a hiker exploring different trails (models), stopping along each trail to take measurements (parameters) before deciding whether to move onto another path.

Using RJMCMC, we can estimate the posterior probability that each covariate influences survival, and obtain model-averaged regression coefficients and survival estimates that incorporate both parameter and model uncertainty.

I won't dive into the mathematical details here but don't worry, you can still follow the analysis without them. Check out Chapter 7 of King et al. [2009; see also Gimenez et al., 2009a] if you're interested in.

6.2.2   Model and NIMBLE implementation

For our example, we'll work with data from 321 encounter histories of white storks ringed as chicks between 1956 and 1970. These data were kindly provided by Jean-Dominique Lebreton. We'll fit a Cormack–Jolly–Seber (CJS) model with time-dependent survival and constant detection probability, as described in Section 4.5. We model the survival probability `phi` as a linear function of our covariates stored in `x`. To handle model selection inside the RJMCMC, we use a trick: for each covariate, we introduce an indicator variable `ksi` that can take the value 1 (covariate is included in the model) or 0 (covariate is excluded). We then combine the regression coefficients `beta` with these indicators into a single vector `betaksi`. This means that if `ksi[j] = 0`, the effect of covariate `j` is zero, effectively removing it from the model. Here's how it works in the NIMBLE code:

```
for (t in 1:(T-1)){
  logit(phi[t]) <- intercept + inprod(betaksi[1:10], x[t, 1:10])
  gamma[1,1,t] <- phi[t]        # Pr(alive t -> alive t+1)
  gamma[1,2,t] <- 1 - phi[t]    # Pr(alive t -> dead t+1)
  gamma[2,1,t] <- 0             # Pr(dead t -> alive t+1)
  gamma[2,2,t] <- 1             # Pr(dead t -> dead t+1)
}
```

For the indicators and regression coefficients:

```
for (j in 1:10){
  betaksi[j] <- beta[j] * ksi[j]
  ksi[j] ~ dbern(psi) # indicator variable associated with betaj
}
psi ~ dbeta(1, 1) # prior on inclusion probability
for (j in 1:10){
  beta[j] ~ dnorm(0, sd = 1.5) # prior slopes
}
intercept ~ dnorm(0, sd = 1.5) # prior intercept
```

The prior on `psi` controls the overall tendency to include covariates in the model. Setting it to `dbeta(1, 1)` makes all inclusion probabilities equally likely, so the data drive the selection. The priors on `beta` and the `intercept` (not consider in the selection) are weakly informative.

As we saw in Section 2.6.1, NIMBLE allows you to change the default sampler. Here, we're going to do something slightly different and use NIMBLE's `configureRJ()` function to set up RJMCMC for variable selection. The function `configureRJ()` modifies an existing MCMC configuration so that RJMCMC is used to sample certain parameters, allowing covariates to be turned on or off during the run. It uses a univariate normal proposal distribution, and you can adjust the proposal mean and scale. The main arguments are:
- `targetNodes`: The parameters you want to apply variable selection to (e.g. `beta`);
- `indicatorNodes`: The indicator variables paired with the target nodes (e.g. `ksi`);
- `control`: A list for fine-tuning the proposal with `mean` the mean of the proposal distribution (default is 0), `scale` the standard deviation of the proposal (default is 1) and `fixedValue` the value a variable takes when excluded (default is 0).

Here's how we apply it to our white stork example:

```r
## Build the model and configure default MCMC
RJsurvival <- nimbleModel(code = hmm.phirjmcmcp,
                          constants = my.constants,
                          data = my.data,
                          inits = initial.values())

Csurvival <- compileNimble(RJsurvival)

RJsurvivalConf <- configureMCMC(RJsurvival)
RJsurvivalConf$addMonitors('ksi')

# Switch to reversible jump sampling for the betas
configureRJ(conf = RJsurvivalConf,
            targetNodes = 'beta',
            indicatorNodes = 'ksi',
            control = list(mean = 0, scale = 2))
```

We then build, compile, and run the RJMCMC:

```r
survivalMCMC <- buildMCMC(RJsurvivalConf)
CsurvivalMCMC <- compileNimble(survivalMCMC, project = RJsurvival)
samples <- runMCMC(mcmc = CsurvivalMCMC,
                   niter = n.iter,
                   nburnin = n.burnin,
                   nchains = n.chains)
```

When we run this MCMC, `configureRJ()` takes care of proposing moves that either keep a covariate in the model or drop it out by setting its `ksi` value to 0. If a covariate is included (`ksi = 1`), the sampler updates its regression coefficient `beta` using the specified normal proposal distribution. If it's excluded (`ksi = 0`), the coefficient is fixed at zero (our `fixedValue`). Over many iterations, the chain "jumps" between different combinations of covariates, building up posterior probabilities for each one's inclusion. This means we can later summarize which covariates are most likely to affect survival (via their inclusion probabilities), and model-averaged estimates of survival that incorporate the uncertainty about which covariates belong in the model. This is what we do in R in the next section.

6.2.3   Results and interpretation

The RJMCMC run gives us the posterior probabilities of different models where each model corresponds to a specific combination of covariates. To get these probabilities, we first combine the MCMC samples from our two chains:

```r
# Gather two chains
out <- rbind(samples[[1]], samples[[2]])
```

Next, we look at the inclusion indicators `ksi`, which take value 1 if a covariate is in the model and 0 if not. The posterior mean of each `ksi[j]` is simply the estimated probability that covariate j is included:

```r
# Number of covariates
K <- 10
```

```r
# Extract only ksi columns, in order ksi[1], ..., ksi[10]
ksi <- out[, paste0("ksi[", 1:K, "]")]
```

```r
# Posterior inclusion probabilities for each covariate
inclusion_prob <- colMeans(ksi)
inclusion_prob
##  ksi[1]  ksi[2]  ksi[3]  ksi[4]  ksi[5]  ksi[6]  ksi[7]
## 0.02575 0.01720 0.01560 0.96120 0.04430 0.05350 0.02025
##  ksi[8]  ksi[9] ksi[10]
## 0.03360 0.02630 0.02965
```

The posterior inclusion probabilities tell us how likely each covariate is to appear in the "true" model given the data and priors. Here, `ksi[4]` stands out, with an inclusion probability around 0.96. This corresponds to the Kita station.

We can also identify the most probable models visited by the RJMCMC. Each MCMC draw corresponds to a model, which we encode as a binary string (e.g., "0101001001"):

```r
# Encode each model draw as a string of 0s and 1s
model_id <- apply(ksi, 1, paste0, collapse = "")

# Empirical posterior model probabilities
post_model_prob <- sort(prop.table(table(model_id)),
                        decreasing = TRUE)

# Show the top models
head(data.frame(model = names(post_model_prob),
                prob = as.numeric(post_model_prob)), 10)
##             model     prob
## 1   0001000000 0.75470
## 2   0001010000 0.03965
## 3   0001100000 0.02910
## 4   0000000000 0.02310
## 5   0001000001 0.02055
## 6   0001000100 0.01710
## 7   0001000010 0.01660
## 8   1001000000 0.01575
## 9   0001001000 0.01320
## 10  0101000000 0.01045
```

The posterior model probabilities rank the different combinations of covariates, and in our case, the model containing only the rainfall at Kita is clearly dominant.

In conclusion, a very high inclusion probability for Kita (covariate #4) and a dominant model containing only Kita indicate that this station's rainfall is strongly linked to survival, while others are not.

Now turning to inference, we'd like to get model-averaged parameters.

Each covariate's effect is "on" only when its indicator (ksi) is 1. Multiplying beta by ksi zeros out the effect when that covariate is excluded. Summarizing betaksi across draws gives the model-averaged effect on the logit scale:

```r
# Pull posterior draws
beta  <- out[, paste0("beta[", 1:K, "]")]


# Model-averaged coefficients on the logit scale
betaksi <- beta * ksi


# Summaries: mean, 95% credible intervals, and Pr(effect > 0)
df_betaksi <- data.frame(
  mean  = apply(betaksi, 2, mean),
  low  = apply(betaksi, 2, quantile, probs = 2.5/100),
  high  = apply(betaksi, 2, quantile, probs = 97.5/100),
  P_gt0 = apply(betaksi > 0, 2, mean))


round(df_betaksi, 3)
##                 mean      low   high P_gt0
## beta[1]     0.003   0.000 0.000 0.023
## beta[2]     0.000   0.000 0.000 0.009
## beta[3]     0.000   0.000 0.000 0.010
## beta[4]     0.348   0.000 0.545 0.961
## beta[5]    -0.006  -0.123 0.000 0.003
## beta[6]     0.007   0.000 0.144 0.051
## beta[7]    -0.002   0.000 0.000 0.004
## beta[8]     0.005   0.000 0.048 0.028
## beta[9]     0.002   0.000 0.000 0.021
## beta[10]   -0.003  -0.020 0.000 0.003
```

A large P_gt0 (e.g., > 0.95) suggests strong evidence that the covariate increases survival (on the logit scale).

Now let's get model-averaged survival estimates over time. For each

MCMC draw, we build the linear predictor with all covariates using `beta * ksi`, add the intercept, then inverse-logit to get survival `phi`:

```r
invlogit <- function(x) 1 / (1 + exp(-x))

# Table of covariate values
X <- rainfall[1:(16 - 1), 1:K]

# Intercept MCMC draws
intercept <- out[, "intercept"]

# Linear predictor for every time t (rows)
# and every MCMC draw (columns)
eta <- X %*% t(betaksi)
eta <- sweep(eta, 2, intercept, "+")   # add intercept per draw

# Back-transform to survival probabilities
phi_draws <- invlogit(eta)

# Summaries by time
df_phi <- data.frame(
  mean  = apply(phi_draws, 1, mean),
  low   = apply(phi_draws, 1, quantile, probs = 2.5/100),
  high  = apply(phi_draws, 1, quantile, probs = 97.5/100))

# Visualize
ggplot(df_phi, aes(x = 1956:1970, y = mean)) +
  geom_ribbon(aes(ymin = low, ymax = high), alpha = 0.2) +
  geom_line(linewidth = 0.9) +
  geom_point() +
  labs(
    x = "Years (interval start)",
    y = "Estimated survival probability",
    title = "Model-averaged survival with 95% credible intervals"
  )
```

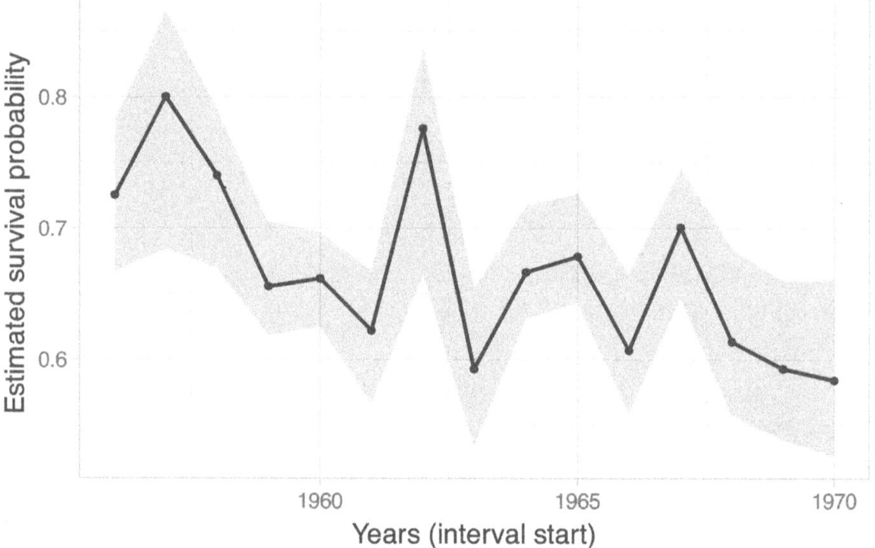

Shaded ribbons show 95% credible intervals; the solid line is the posterior mean of survival, which decreases over time, model-averaged over covariate inclusion.

6.3   Uncertainty in age

6.3.1   Motivation

Survival and other demographic parameters often differ between young and older animals, so if we blur age classes we can end up telling the wrong story about population dynamics. In Section 4.8.5, we incorporated age as a covariate in capture-recapture models, assuming it was perfectly known. With non-invasive genetics (Figure 4.1), though, age is not observed: a DNA profile can identify who was there, but not how old they were. If we ignore that uncertainty, survival estimates can be biased. Inspired by what we saw in Section 5.7, we will see here how HMM capture-recapture models can handle this kind of problem by treating age as a hidden state and separating it from what we actually observe in the field.

A neat example comes from Apennine brown bears in Gervasi et al. [2017; see also Gowan et al., 2021]. Because you cannot age a bear from its DNA, the authors paired genetics with field information to classify detections into two broad age classes, cubs (<1 year) and adults (≥ 1 year). To do so, they used two criteria. First, classify a newly detected bear as a cub in that year if it was sampled at least once on the same date and site as a known adult female and they shared ≥ 1 allele at every locus (mother-offspring consistent). That's the lenient criterion (P1). Second, apply a stricter criterion (P2): make the same mother–offspring call, but require two or more such co-sampling occasions with the same female, again on the same date and site each time.

Why two criteria? Because there's a trade-off. P1 is better at catching real cubs (more sensitive) but risks misclassifying some adults as cubs; P2 is more conservative about calling "cub", so it reduces false cubs but misses some true ones.

Instead of pretending these criteria are perfect, the authors model the misclassification explicitly: age is hidden, "cub/adult by P1/P2" is what we observe, and their model estimates both classification accuracy and age-specific survival at the same time. That way, uncertainty is propagated into the demographic estimates rather than swept under the rug.

Here, I show how to encode that logic in an HMM with NIMBLE: define the hidden age states, write the transition (survival) and detection components, and add a classification step that links hidden age to the observed P1/P2 outcomes. We'll also see how a small subset of known-age individuals (e.g., live-trapped or aged beforehand) anchors the classification step. The payoff is a set of age-specific survival estimates that are robust to imperfect age assignment, which is exactly what we need for sound ecological inference and management.

6.3.2 Model and NIMBLE implementation

To estimate bear survival, we have at our disposal a 12-year time series of non-invasive genetic sampling data, collected between 2003 and 2014. These data were kindly provided by Vincenzo Gervasi. To build up our model, we first define the states and observations:

- States
    - alive as cub (C)
    - alive as adult (A)
    - dead (D)
- Observations
    - not detected (1)
    - detected and classified as cub by both criteria (2)
    - detected and classified as cub by P1 only (3)
    - detected and classified as adult by both criteria (4)

Now we turn to writing our model.

We start with the vector of initial states. Unknown-age individuals enter as a mixture of cubs and adults:

$$\delta = \begin{pmatrix} \overset{z_t = C}{\pi} & \overset{z_t = A}{1-\pi} & \overset{z_t = D}{0} \end{pmatrix}$$

where π is the probability of being alive as a cub, and $1 - \pi$ is the probability of being alive as an adult. For bears with known age at first detection, we fix the initial state accordingly. In NIMBLE we implement this with an individual-specific δ_i using `equals()` to pin known ages:

```
# Initial state
# age[i] == 0: unknown -> use delta: pi, 1 - pi
# age[i] == 1: known cub -> force delta to [1, 0, 0]
# age[i] == 2: known adult -> force delta to [0, 1, 0]
for (i in 1:N){
  # Pr(C)
  delta[1, i] <- equals(age[i], 0) * pi + equals(age[i], 1)
  # Pr(A)
  delta[2, i] <- equals(age[i], 0) * (1 - pi) + equals(age[i], 2)
  delta[3, i] <- 0
}
```

We proceed with the transition matrix that contains the survival probabilities. A cub either survives and becomes adult next year (ϕ_C), or dies; adults either survive in the same class (ϕ_A) or die:

$$\Gamma = \begin{array}{c} \quad z_t = C \quad z_t = A \quad z_t = D \\ \begin{pmatrix} 0 & \phi_C & 1 - \phi_C \\ 0 & \phi_A & 1 - \phi_A \\ 0 & 0 & 1 \end{pmatrix} \begin{array}{l} z_{t-1} = C \\ z_{t-1} = A \\ z_{t-1} = D \end{array} \end{array}$$

Following the paper, adult survival is sex-specific; cub survival is common to both sexes:

```
# Transition matrix
for (i in 1:N){
  gamma[1,1,i] <- 0
  gamma[1,2,i] <- phiC
  gamma[1,3,i] <- 1 - phiC
  gamma[2,1,i] <- 0
  gamma[2,2,i] <- phiA[sex[i]] # sex-spec adult surv (1=f, 2=m)
  gamma[2,3,i] <- 1 - phiA[sex[i]]
  gamma[3,1,i] <- 0
  gamma[3,2,i] <- 0
  gamma[3,3,i] <- 1
}
```

Last step is about the observations, which arise in two steps: detection then classification. For the detection matrix, we need to introduce intermediate observations, say y' for not detected (1), detected as cub (2) and detected as adult (2):

$$\Omega_1 = \begin{array}{c} \quad y'_t = 1 \quad y'_t = 2 \quad y'_t = 3 \\ \begin{pmatrix} 1 - p_C & p_C & 0 \\ 1 - p_A & 0 & 1 - p_A \\ 1 & 0 & 0 \end{pmatrix} \begin{array}{l} z_t = C \\ z_t = A \\ z_t = D \end{array} \end{array}$$

where p_C is detection for cubs, and p_A that for adults. Then the classification matrix is:

$$\Omega_2 = \begin{array}{cc} & \begin{array}{cccc} y_t = 1 & y_t = 2 & y_t = 3 & y_t = 4 \end{array} \\ \begin{array}{c} \\ \\ \\ \end{array} & \left(\begin{array}{cccc} 1 & 0 & 0 & 0 \\ 0 & \beta_{C,CC} & \beta_{C,CA} & 1 - \beta_{C,CC} - \beta_{C,CA} \\ 0 & \beta_{A,CC} & \beta_{A,CA} & 1 - \beta_{A,CC} - \beta_{A,CA} \end{array} \right) & \begin{array}{c} y'_t = 1 \\ y'_t = 2 \\ y'_t = 3 \end{array} \end{array}$$

where $\beta_{C,CC}$ is the probability that, being a cub, and individual is classified as cub with both criteria, $\beta_{C,CA}$ is the probability that, being a cub, an individual is classified as cub with P1 and as adult with P2, $\beta_{A,CC}$ is the probability that, being an adult, an individual is classified as cub with both criteria, $\beta_{A,CA}$ is the probability that, being an adult, an individual is classified as cub with P1 and as adult with P2.

The observation matrix is the product of the two detection and classification matrices $\Omega_1\Omega_2$:

$$\Omega = \begin{array}{cc} & \begin{array}{cccc} y_t = 1 & y_t = 2 & y_t = 3 & y_t = 4 \end{array} \\ \begin{array}{c} \\ \\ \\ \end{array} & \left(\begin{array}{cccc} 1 - p_C & p_C\beta_{C,CC} & p_C\beta_{C,CA} & p_C(1 - \beta_{C,CC} - \beta_{C,CA}) \\ 1 - p_A & p_A\beta_{A,CC} & p_A\beta_{A,CA} & p_A(1 - \beta_{A,CC} - \beta_{A,CA}) \\ 1 & 0 & 0 & 0 \end{array} \right) & \begin{array}{c} z_t = 1 \\ z_t = 2 \\ z_t = 3 \end{array} \end{array}$$

In code, we write $\Omega_1\Omega_2$ directly, as this is faster than doing the matrix multiplication at each MCMC iteration. For simplicity we use a single detection probability p for both ages, but you can let p vary by age, sex, year or sampling method, as in the paper:

```r
# Observation matrix
omega[1,1] <- 1 - p
omega[1,2] <- p * betaCCC
omega[1,3] <- p * betaCCA
omega[1,4] <- p * (1 - betaCCC - betaCCA)

omega[2,1] <- 1 - p
omega[2,2] <- p * betaACC
omega[2,3] <- p * betaACA
omega[2,4] <- p * (1 - betaACC - betaACA)
```

```
omega[3,1] <- 1
omega[3,2] <- 0
omega[3,3] <- 0
omega[3,4] <- 0
```

At first detection, the event "not detected" is impossible by construction (we condition on the first detection). We therefore use an initial observation matrix Ω_{init} with p set to 1 (not shown).

The likelihood and priors are handled as usual.

6.3.3 Results and interpretation

Here are the raw results:

```
##              mean   sd 2.5%  50% 97.5% Rhat n.eff
## betaACA 0.05 0.02 0.02 0.04  0.08 1.00  4114
## betaACC 0.02 0.01 0.00 0.02  0.05 1.00  4211
## betaCCA 0.57 0.12 0.33 0.58  0.79 1.00   966
## betaCCC 0.30 0.10 0.13 0.29  0.50 1.00   998
## p       0.71 0.04 0.63 0.71  0.79 1.00  3329
## phiA[1] 0.88 0.03 0.80 0.88  0.94 1.00  4195
## phiA[2] 0.82 0.06 0.69 0.82  0.92 1.00  3643
## phiC    0.53 0.13 0.28 0.53  0.78 1.00  3563
## pi      0.27 0.09 0.12 0.26  0.49 1.02  1416
```

which we can arrange as in Table 6.1 to compare them with the results obtained by Gervasi et al. [2017].

Intervals in Table 6.1 are 95% credible intervals for NIMBLE and confidence intervals for Gervasi et al. [2017]. Detection in Gervasi et al. [2017] varies by design and year; I report their overall average.

Our NIMBLE fit reproduces the main biological signals reported for the Apennine brown bear. Survival shows the expected ordering, that cubs « adults, and adult females > adult males – with estimates that closely match the published analysis. The credible/confidence intervals overlap broadly, indicating good agreement.

TABLE 6.1: Comparison of parameter estimates for the same HMM to account for age uncertainty: our NIMBLE fit vs. Gervasi et al. (2017, Table 3).

Parameter	NIMBLE	Gervasi et al. (2017)
ϕ_C (cub survival)	0.53 (0.28–0.78)	0.51 (0.22–0.79)
$\phi_{A,f}$ (adult female)	0.88 (0.80–0.94)	0.92 (0.87–0.95)
$\phi_{A,m}$ (adult male)	0.82 (0.69–0.92)	0.85 (0.76–0.91)
p (detection)	0.71 (0.63–0.79)	≈ 0.71 (0.62–0.79)
π (cub at first detection)	0.27 (0.12–0.49)	0.062 (0.007–0.396)
$C_{c,cc}$ (cub by both given cub)	0.30 (0.13–0.50)	0.40 (0.15–0.72)
$C_{c,ca}$ (cub by P1 only given cub)	0.57 (0.33–0.79)	0.60 (0.28–0.85)
$C_{a,cc}$ (cub by both given adult)	0.02 (0.00–0.05)	0.067 (0.024–0.174)
$C_{a,ca}$ (cub by P1 only given adult)	0.05 (0.02–0.08)	0.191 (0.110–0.309)

Our detection estimate aligns with the study's average detection, even though their analysis lets detection vary by design and year rather than assuming a single constant value.

For age classification (linking the hidden age to P1/P2 outcomes), mapping our parameters to theirs, we found lower adult-as-cub misclassification than the paper (and thus a higher probability that detected adults are correctly called "adult by both"), which still matches their qualitative take: the stricter criterion (P2) performs very well for adults.

The main divergence is in the initial state mix. Intervals overlap, but the means differ. A likely explanation is model structure: we used a single, constant detection probability, while the published analysis allows detection to vary by sampling design and year – and detectability does vary meaningfully across methods and years. Collapsing this heterogeneity into one p can nudge the mixture at first detection toward a higher cub fraction to explain patterns that the richer detection model attributes to design or year effects.

In conclusion, we managed to recover the core findings of the published study (age- and sex-specific survival and average detection). Small discrepancies appear mainly in adult misclassification rates and π. To mirror Gervasi et al. [2017] even more closely, the natural next step is to let p vary by sampling design/year (an information I did not have), as in their best supported model.

6.4 Uncertainty in sex

6.4.1 Motivation

Many species show sex differences in demographic parameters, so mislabeling sex or discarding uncertain cases can bias the very patterns we want to estimate. In monomorphic birds for example, field sexing relies on behavior (e.g., copulation posture, courtship displays, relative body size), and those clues vary in reliability; in the Audouin's gull (*Larus audouinii*) study that motivates this section, almost 80% of individuals were never sexed in the field, making "known-sex only" analyses both wasteful and biased toward higher survival.

Pradel et al. [2008; see also Genovart et al., 2012] turned this problem into a HMM inference task: treat sex as a hidden state (male/female/dead) and model the sex-assignment process itself. The model developed by Pradel, Genovart and colleagues separates (i) the probability of attempting a sex judgment (with a time trend, because sexing effort increased) from (ii) criterion-specific accuracy for four behavioral clues (copulation, courtship feeding, begging for food, body size).

In brief, the message is: when field classifications are uncertain, exploit them rather than filter them out, and use HMM to couple biological states to imperfect observations and propagate classification uncertainty into survival.

That philosophy mirrors the age-uncertainty Section 6.3 we just completed: there, age was hidden and P1/P2 provided noisy age labels; here, sex is hidden and multiple behavioral criteria provide noisy sex labels.

6.4.2 Model and NIMBLE implementation

To estimate Audouin's gull survival, we have at our disposal a dataset with 4093 marked birds from a colony at the Ebro Delta (Spain),

collected over 10 years. These data were kindly provided by Meritxell Genovart and Roger Pradel. To build up our model, we first define
the states and observations:

- States
    - alive as male (M)
    - alive as female (F)
    - dead (D)
- Observations
    - not seen (1)
    - judged from copulation to be male (2)
    - judged from begging food to be male (3)
    - judged from courtship feeding to be male (4)
    - judged from body size to be male (5)
    - judged from copulation to be female (6)
    - judged from begging food to be female (7)
    - judged from courtship feeding to be female (8)
    - judged from body size to be female (9)
    - not judged (10)

Now we turn to writing our model.

We start with the vector of initial states. Unknown-sex individuals enter
as a mixture of males and females with probabilities π and $1 - \pi$:

$$\delta = \begin{pmatrix} z_t = M & z_t = F & z_t = D \\ \pi & 1 - \pi & 0 \end{pmatrix}$$

We proceed with the transition matrix that contains the survival probabilities. A male either survives next year as a male (ϕ_M), or dies; the
same happens for females (with probability ϕ_F):

$$\Gamma = \begin{pmatrix} z_t = M & z_t = F & z_t = D \\ \phi_M & 0 & 1 - \phi_M \\ 0 & \phi_F & 1 - \phi_F \\ 0 & 0 & 1 \end{pmatrix} \begin{matrix} \\ z_{t-1} = M \\ z_{t-1} = F \\ z_{t-1} = D \end{matrix}$$

The vector of initial states and the matrix of transition probabilities are
written as usual in NIMBLE:

```
delta[1] <- pi
delta[2] <- 1 - pi
delta[3] <- 0

gamma[1,1] <- phiM
gamma[1,2] <- 0
gamma[1,3] <- 1 - phiM
gamma[2,1] <- 0
gamma[2,2] <- phiF
gamma[2,3] <- 1 - phiF
gamma[3,1] <- 0
gamma[3,2] <- 0
gamma[3,3] <- 1
```

Last step is the observation matrix, which we write as follows (its transposed for convenience):

$$
\Omega' = \begin{array}{c}
\begin{array}{ccc}
z_t = M & z_t = F & z_t = D
\end{array} \\
\left(\begin{array}{ccc}
1-p & 1-p & 1 \\
pe(1-m_4)m_1 x_1 & pe(1-m_4)m_1(1-x_1) & 0 \\
pe(1-m_4)m_2 x_2 & pe(1-m_4)m_2(1-x_2) & 0 \\
pe(1-m_4)m_3 x_3 & pe(1-m_4)m_3(1-x_3) & 0 \\
pem_4 x_4 & pem_4(1-x_4) & 0 \\
pe(1-m_4)m_1(1-x_1) & pe(1-m_4)m_1 x_1 & 0 \\
pe(1-m_4)m_2(1-x_2) & pe(1-m_4)m_2 x_2 & 0 \\
pe(1-m_4)m_3(1-x_3) & pe(1-m_4)m_3 x_3 & 0 \\
pem_4(1-x_4) & pem_4 x_4 & 0 \\
p(1-e) & p(1-e) & 0
\end{array}\right)
\begin{array}{c}
y_t = 1 \\
y_t = 2 \\
y_t = 3 \\
y_t = 4 \\
y_t = 5 \\
y_t = 6 \\
y_t = 7 \\
y_t = 8 \\
y_t = 9 \\
y_t = 10
\end{array}
\end{array}
$$

This matrix is a bit more complex than what we have encountered before, and needs some explanations. When an individual is encountered (with probability p), there is some chance that it will be sexed (e) based on its general appearance (body size: probability m_4) or its behavior (probability $1 - m_4$). When behavior is used, there are three different criteria (copulation, begging for food and courtship feeding), which occur with probabilities m_1, m_2, and m_3, respectively. Finally, each criterion has its own reliability (probabilities x_1, x_2, x_3, x_4, where x_4 is the reliability of the body-size criterion).

I admit that this matrix representation can be difficult to grasp at once. It can be easier to split the observation matrix in several steps, like encountering, sexing, body size sexing, sex determination and correctness, then to sketch a decision tree as in Figure 2 of Genovart et al. [2012] (see also appendix S1 in that same paper).

In code, this gives:

```
omega[1,1]  <- 1 - p
omega[1,2]  <- p * e * (1 - m[4]) * m[1] * x[1]
omega[1,3]  <- p * e * (1 - m[4]) * m[2] * x[2]
omega[1,4]  <- p * e * (1 - m[4]) * m[3] * x[3]
omega[1,5]  <- p * e * m[4] * x[4]
omega[1,6]  <- p * e * (1 - m[4]) * m[1] * (1 - x[1])
omega[1,7]  <- p * e * (1 - m[4]) * m[2] * (1 - x[2])
omega[1,8]  <- p * e * (1 - m[4]) * m[3] * (1 - x[3])
omega[1,9]  <- p * e * m[4] * (1 - x[4])
omega[1,10] <- p * (1 - e)

omega[2,1]  <- 1 - p
omega[2,2]  <- p * e * (1 - m[4]) * m[1] * (1 - x[1])
omega[2,3]  <- p * e * (1 - m[4]) * m[2] * (1 - x[2])
omega[2,4]  <- p * e * (1 - m[4]) * m[3] * (1 - x[3])
omega[2,5]  <- p * e * m[4] * (1 - x[4])
omega[2,6]  <- p * e * (1 - m[4]) * m[1] * x[1]
omega[2,7]  <- p * e * (1 - m[4]) * m[2] * x[2]
omega[2,8]  <- p * e * (1 - m[4]) * m[3] * x[3]
omega[2,9]  <- p * e * m[4] * x[4]
omega[2,10] <- p * (1 - e)

omega[3,1]  <- 1
omega[3,2]  <- 0
omega[3,3]  <- 0
omega[3,4]  <- 0
```

```
omega[3,5] <- 0
omega[3,6] <- 0
omega[3,7] <- 0
omega[3,8] <- 0
omega[3,9] <- 0
omega[3,10] <- 0
```

At first detection, the observation "not detected" is impossible by construction (we condition on the first detection). We therefore use an initial observation matrix Ω_{init} with p set to 1 (not shown).

The likelihood and priors are handled as usual.

6.4.3 Results and interpretation

Here are the raw results:

```
##          mean    sd 2.5%   50% 97.5% Rhat n.eff
## e        0.07 0.00 0.07 0.07  0.07 1.01   589
## m[1]     0.29 0.02 0.26 0.29  0.32 1.00   913
## m[2]     0.60 0.02 0.56 0.60  0.63 1.00   892
## m[3]     0.11 0.01 0.09 0.11  0.14 1.00   817
## m[4]     0.17 0.01 0.14 0.17  0.19 1.01   949
## p        0.65 0.00 0.64 0.65  0.66 1.00   394
## phiF     0.90 0.01 0.89 0.90  0.91 1.01   205
## phiM     0.89 0.01 0.88 0.89  0.90 1.01   279
## pi       0.52 0.01 0.50 0.52  0.55 1.00   291
## x[1]     0.52 0.03 0.45 0.52  0.58 1.01   915
## x[2]     0.55 0.02 0.50 0.55  0.60 1.01   725
## x[3]     0.54 0.05 0.44 0.55  0.65 1.00   927
## x[4]     0.58 0.04 0.50 0.58  0.67 1.00   753
```

which we can arrange in as in Table 6.2 to compare them with the results obtained by Pradel et al. [2008].

Intervals in Table 6.2 are 95% credible intervals for NIMBLE and confidence intervals for Pradel et al. [2008]. Our error rates are 1 - x[i]. Pradel's m's were provided by the main author of the paper himself.

TABLE 6.2: Comparison of parameter estimates for the same HMM to account for sex uncertainty: our NIMBLE fit vs. Pradel et al. (2008, Table 5, model B no genetically sexed anchors).

Parameter	NIMBLE	Pradel et al. (2008)
μ (proportion male)	0.52 (0.50–0.55)	0.53 (SE 0.03)
ϕ_F (female survival)	0.90 (0.89–0.91)	0.91 (SE 0.01)
ϕ_M (male survival)	0.89 (0.88–0.90)	0.86 (SE 0.01)
m_1 (Copulation)	0.29 (0.26–0.32)	0.29
m_2 (Begging)	0.60 (0.56–0.63)	0.60
m_3 (Courtship feeding)	0.11 (0.09–0.14)	0.11
m_4 (Body size)	0.17 (0.14–0.19)	time-varying
Error (Copulation)	0.48 (0.42–0.55)	0.06 (SE 0.04)
Error (Begging)	0.45 (0.40–0.50)	0.06 (SE 0.03)
Error (Courtship feeding)	0.46 (0.35–0.56)	0.00 (SE 0.16)
Error (Body size)	0.42 (0.33–0.50)	0.09 (SE 0.07)

Note that m4 was time-varying; the estimates were for $t = 1, \ldots, 10$: 0.001, 0.002, 0.004, 0.007, 0.014, 0.026, 0.047, 0.084, 0.146, 0.242.

Our NIMBLE fit recovers the main biological signals reported for the Audouin's gulls analysis. Survival shows the expected ordering – adult females $\geq$ adult males – with broadly overlapping intervals and means that are close to the published model. In our run, the female–male gap is a bit smaller (0.90 vs 0.89), but the direction matches the paper's message (higher female survival).

We used a single, constant detection probability and obtained $p \approx 0.65$, which sits near the middle of the year-specific values obtained by the authors (not shown in the paper).

The estimated probabilities of which clue is used align closely with the published analysis for the three "behavioural" criteria: $m_1 \approx 0.29$ (copulation), $m_2 \approx 0.60$ (begging), $m_3 \approx 0.11$ (courtship feeding). The paper lets body size use vary over time; our constant estimate ($m_4 \approx 0.17$) should be read as a pooled average across years—consistent with their finding of a low but increasing use of this criterion.

The main discrepancy lies in the error rates by criterion. The published analysis finds low misclassification (roughly 5–10% for most criteria, with a very imprecise estimate for courtship feeding). Our fit suggests substantially higher error across criteria. That discrepancy is where we expect sensitivity when there are no known-sex anchors (and some components are simplified). Without anchors, the model can admit dual solutions (swapping sex labels and flipping error probabilities) that fit nearly equally well; small modeling choices or initial values can tip the chain toward one mode, see Pradel et al. [2008].

Our estimate of the proportion of males at first encounter ($\mu \approx 0.52$) agrees closely with the published result, reinforcing that the state mix is well constrained by the data even when sex is uncertain.

To conclude, the sex-uncertainty HMM retrieves the core demographic pattern (female $\geq$ male survival) and matches the paper on overall detectability and the use of criteria. To mirror the published results more closely, one could add a small anchor set of genetically sexed individuals [an information used in Pradel et al., 2008, but that I did not have].

7

Addressing model lack of fit

7.1   Introduction

Capture-recapture models rely on assumptions that must hold for inference to be reliable. In Section 4.7, we saw what could go wrong and how to diagnose those issues. In this chapter, we turn to remedies and show how to fix them by building models that explicitly accommodate lack of fit. We focus on three topics: trap-dependence (detection today affects detection tomorrow), memory in the state process (the Markov assumption is too short – transitions depend on more than the current state), and individual heterogeneity (persistent differences among animals).

7.2   Accounting for trap-dependence

7.2.1   Motivation

Capture-recapture inference can go sideways when detection depends on past capture. If animals become trap-shy (avoid traps) or trap-happy (seek them) after being caught, the independence assumption breaks, and survival or detection can be biased if we ignore it. Tests for this trap-dependence exist, see Section 4.7, and surveys show it's common across taxa – roughly 70% of reviewed studies reported evidence for it [Pradel and Sanz-Aguilar, 2012].

The fix is to encode the behavioral response. Following Pradel and Sanz-Aguilar [2012], we treat trap-awareness as a (possibly temporary) state: right after a capture, individuals move to a "trap-aware" state with its

DOI: 10.1201/9781003244097-7

own detection probability, so that detection at time $t + 1$ can differ for animals caught vs. not caught at t. This HMM formulation avoids awkward data splitting, and plays nicely with age or covariates. Importantly, in the authors' example, ignoring trap-dependence underestimated survival, illustrating why modeling trap-dependence matters.

Here, I adopt that logic in NIMBLE: define a trap-aware state, let detection depend on it, and keep the rest of the HMM machinery unchanged.

7.2.2 Model and NIMBLE implementation

For our example, we'll work with data from 519 encounter histories of Cory's shearwaters (*Calonectris diomedea*) marked as adults between 2001 and 2008 at Pantaleu Islet (Balearic Archipelago, Spain). These data were kindly provided by Ana Sanz-Aguilar. If you perform goodness-of-fit tests on this dataset, as in Section 4.7, you will end up with a significant trap-dependence effect (as well as a transience effect). To build up our model and explicitly account for this lack of fit, we first define the states and observations:

- States
  - A for "trap-aware", the original state of an individual when it is first released
  - U for "trap-unaware", which follows any occasion where it is not captured
  - D for dead

- Observations
  - not detected (1)
  - detected (2)

Now we turn to writing our model. We start with the vector of initial states. Individuals enter as trap-aware with probability 1:

$$\delta = \begin{pmatrix} \overset{z_t = A}{1} & \overset{z_t = U}{0} & \overset{z_t = D}{0} \end{pmatrix}$$

We proceed with the transition matrix:

$$
\Gamma = \begin{array}{c}
\begin{array}{ccc} z_t = A & z_t = U & z_t = D \end{array} \\
\left(\begin{array}{ccc}
\phi p' & \phi(1 - p') & 1 - \phi \\
\phi p & \phi(1 - p) & 1 - \phi \\
0 & 0 & 1
\end{array} \right)
\end{array}
\begin{array}{c}
z_{t-1} = A \\
z_{t-1} = U \\
z_{t-1} = D
\end{array}
$$

You might have noticed something unusual in that Γ contains both the survival and detection probabilities. Why is that? We track three hidden states from the end of session t to $t + 1$: A = trap-aware (the bird was just caught at t), U = trap-unaware (not caught at t), and D = dead. Between sessions, everyone either survives with probability ϕ or dies with $1 - \phi$ (which moves them to D, an absorbing state). If alive at $t + 1$, the awareness flag for the next session is set by whether the bird gets caught at $t + 1$: those caught will be A next time, those not caught will be U. Because capture probabilities can differ by current awareness, we allow p' for birds that were A at t and p for birds that were U at t. If $p' > p$ you have trap-happy behavior (recently caught birds are easier to catch next time); if $p' < p$, it's trap-shy. This "survive -> update awareness via capture" factorization is what the transition matrix formalizes.

Following the paper, adult survival is age-specific to account for a transient effect, see Section 4.7:

```
# Transition matrix
  for (i in 1:N){
    for (t in first[i]:(K-1)){
      phi[i,t] <- beta[age[i,t]] # beta1 = phi1, beta2 = phi2+
      gamma[1,1,i,t] <- phi[i,t] * pprime
      gamma[1,2,i,t] <- phi[i,t] * (1 - pprime)
      gamma[1,3,i,t] <- 1 - phi[i,t]
      gamma[2,1,i,t] <- phi[i,t] * p
      gamma[2,2,i,t] <- phi[i,t] * (1 - p)
      gamma[2,3,i,t] <- 1 - phi[i,t]
      gamma[3,1,i,t] <- 0
      gamma[3,2,i,t] <- 0
```

```
    gamma[3,3,i,t] <- 1
  }
}
```

In the code above, age is passed in the data and created as follows:

```
# Age effect via individual x time cov and nested indexing
# to distinguish survival over interval after first detection
# from survival afterwards:
age <- matrix(NA, nrow = nrow(y), ncol = ncol(y) - 1)
for (i in 1:nrow(age)){
  for (j in 1:ncol(age)){
    if (j == first[i]) age[i,j] <- 1 # age = 1
    if (j > first[i]) age[i,j] <- 2  # age > 1
  }
}
```

Last step is the observation matrix:

$$\Omega = \begin{array}{c} \\ \end{array} \overset{\displaystyle y_t = 1 \quad y_t = 2}{\left(\begin{array}{cc} 0 & 1 \\ 1 & 0 \\ 1 & 0 \end{array} \right)} \begin{array}{l} z_t = A \\ z_t = U \\ z_t = D \end{array}$$

If an animal is trap-aware at t, that means that it has just been captured. If it is trap unaware or dead, it has not been captured during
this session. In code, we write:

```
# Observation matrix
  omega[1,1] <- 0
  omega[1,2] <- 1
  omega[2,1] <- 1
  omega[2,2] <- 0
  omega[3,1] <- 1
  omega[3,2] <- 0
```

The likelihood and priors are handled as usual.

7.2.3 Results and interpretation

Here are the raw results with trap-dependence:

```
##            mean   sd 2.5%  50% 97.5% Rhat n.eff
## p          0.46 0.06 0.34 0.46  0.56 1.10   580
## phi1       0.76 0.02 0.71 0.76  0.80 1.03  1892
## phi2       0.87 0.02 0.84 0.87  0.90 1.08   940
## pprime     0.79 0.02 0.75 0.79  0.82 1.06  1154
```

The two detection probabilities, depending on whether a bird was previously caught pprime or not p, differ clearly, providing evidence of trap-happiness. Our survival estimates are very similar to those obtained by Pradel and Sanz-Aguilar [2012] who found $\phi_1 = 0.77\ (0.70, 0.82)$ and $\phi_2 = 0.87\ (0.82, 0.90)$ (check out their Table 1). Interestingly, when we fit the same model by ignoring trap-dependence, we get the following results:

```
##          mean   sd 2.5%  50% 97.5% Rhat n.eff
## p        0.78 0.02 0.75 0.78  0.81 1.00   489
## phi1     0.74 0.02 0.70 0.74  0.79 1.00   615
## phi2     0.84 0.01 0.82 0.84  0.87 1.01   569
```

When compared the our previous estimates, we see that ignoring trap-dependence leads to an underestimation of survival, either ϕ_1 the transient survival or ϕ_2 the resident survival.

The approach described here can be conveniently extended to several sites or states, see Pradel and Sanz-Aguilar [2012].

7.3 Allowing your Markov models to remember

7.3.1 Motivation

How do we make our models remember? So far, the dynamics of our states have been first-order Markov: the state an animal moves to next depends only on the state it occupies now. But if we think of states as sites, transitions are dispersal or migration, and many species show site fidelity or directional return that reflects where they were several steps back.

To relax the first-order assumption, we move to a second-order Markov description, and what happens at $t + 1$ depends on the site at t and at $t - 1$. This idea, often called a memory model, was introduced in multisite capture-recapture by Hestbeck et al. [1991] and Brownie et al. [1993], then formulated cleanly within the HMM framework by Pradel [2005] and Rouan et al. [2009].

To illustrate the memory model, we will use the geese data introduced in Chapter 5. As we saw in Section 5.3.3, there is a strong positive association between the site where an animal was last seen and the site where it will next be seen. This suggests that a first-order model is too simple and that transitions at $t + 1$ should not only depend on t but also $t - 1$.

7.3.2 Model and NIMBLE implementation

As a reminder, the two main ingredients for a HMM capturing dispersal between 2 sites are a transition matrix:

$$\Gamma = \begin{pmatrix} \phi^A(1 - \psi^{AB}) & \phi^A\psi^{AB} & 1 - \phi^A \\ \phi^B\psi^{BA} & \phi^B(1 - \psi^{BA}) & 1 - \phi^B \\ 0 & 0 & 1 \end{pmatrix} \begin{matrix} z_{t-1} = A \\ z_{t-1} = B \\ z_{t-1} = D \end{matrix}$$

$$\begin{matrix} z_t = A & z_t = B & z_t = D \end{matrix}$$

and an observation matrix:

$$\Omega = \begin{pmatrix} 1 - p^A & p^A & 0 \\ 1 - p^B & 0 & p^B \\ 1 & 0 & 0 \end{pmatrix} \begin{matrix} z_t = A \\ z_t = B \\ z_t = D \end{matrix}$$

$$\begin{matrix} y_t = 1 & y_t = 2 & y_t = 3 \end{matrix}$$

From here, how to come up with a HMM formulation of the memory model? We need to keep track of the sites previously visited, and to do so, the trick is to consider states as being pairs of sites occupied:

- States
    - AA is for alive in site A at t and alive in site A at $t - 1$
    - AB is for alive in site A at t and alive in site B at $t - 1$
    - BA is for alive in site B at t and alive in site A at $t - 1$
    - BB is for alive in site B at t and alive in site B at $t - 1$
    - D is for dead
- Observations
    - 1 not captured
    - 2 captured at site A
    - 3 captured at site B

Now we turn to writing our model. We start with the vector of initial states. Individuals enter the study with probabilities:

$$\delta = \begin{pmatrix} \overset{z_t = AA}{\pi^{AA}} & \overset{z_t = AB}{\pi^{AB}} & \overset{z_t = BA}{\pi^{BA}} & \overset{z_t = BB}{\pi^{BB}} & \overset{z_t = D}{0} \end{pmatrix}$$

where π^{ij} is the probability of being alive at site j when first captured and site i before, and $\pi^{BB} = 1 - (\pi^{AA} + \pi^{AB} + \pi^{BA})$.

We proceed with the transition matrix that contains the survival and movement probabilities:

$$\Gamma = \begin{pmatrix} \overset{z_t = AA}{\phi^{AAA}} & \overset{z_t = AB}{\phi^{AAB}} & \overset{z_t = BA}{0} & \overset{z_t = BB}{0} & \overset{z_t = D}{1 - \phi^{AAA} - \phi^{AAB}} \\ 0 & 0 & \phi^{ABA} & \phi^{ABB} & 1 - \phi^{ABA} - \phi^{ABB} \\ \phi^{BAA} & \phi^{BAB} & 0 & 0 & 1 - \phi^{BAA} - \phi^{BAB} \\ 0 & 0 & \phi^{BBA} & \phi^{BBB} & 1 - \phi^{BBA} - \phi^{BBB} \\ 0 & 0 & 0 & 0 & 1 \end{pmatrix} \begin{matrix} z_{t-1} = AA \\ z_{t-1} = AB \\ z_{t-1} = BA \\ z_{t-1} = BB \\ z_{t-1} = D \end{matrix}$$

where ϕ^{ijk} is the probability to be in site k at time $t + 1$ for an individual present in site j at t and in site i at $t - 1$.

An alternate parameterization for Γ is:

$$\Gamma = \begin{pmatrix} \overset{z_t = AA}{\phi\psi^{AAA}} & \overset{z_t = AB}{\phi(1 - \psi^{AAA})} & \overset{z_t = BA}{0} & \overset{z_t = BB}{0} & \overset{z_t = D}{1 - \phi} \\ 0 & 0 & \phi(1 - \psi^{ABB}) & \phi\psi^{ABB} & 1 - \phi \\ \phi\psi^{BAA} & \phi(1 - \psi^{BAA}) & 0 & 0 & 1 - \phi \\ 0 & 0 & \phi(1 - \psi^{BBB}) & \phi\psi^{BBB} & 1 - \phi \\ 0 & 0 & 0 & 0 & 1 \end{pmatrix} \begin{matrix} z_{t-1} = AA \\ z_{t-1} = AB \\ z_{t-1} = BA \\ z_{t-1} = BB \\ z_{t-1} = D \end{matrix}$$

in which ϕ is the probability of surviving from one occasion to the next, and ψ_{ijj} is the probability an animal stays at the same site j given that it was at site i on the previous occasion.

Last step is the observation matrix, which we write as follows:

$$\Omega = \begin{array}{c} \\ \\ \\ \\ \\ \end{array} \overset{\begin{array}{ccc} y_t = 1 & y_t = 2 & y_t = 3 \end{array}}{\begin{pmatrix} 1 - p^A & p^A & 0 \\ 1 - p^B & 0 & p^B \\ 1 - p^A & p^A & 0 \\ 1 - p^B & 0 & p^B \\ 1 & 0 & 0 \end{pmatrix}} \begin{array}{l} z_t = AA \\ z_t = AB \\ z_t = BA \\ z_t = BB \\ z_t = D \end{array}$$

To implement this model in NIMBLE, we're going to use `nimbleEcology`, which makes our life much easier as we do not have to initialize the latent states z – remember the likelihood is marginalized, see Sections 3.8 and 3.8.3.2. Specifically, we use the function `dHMM()` that implements the distribution of a HMM with constant parameters. We also use a Dirichlet prior to ensure that the initial state probabilities and the survival-movement probabilities sum up to 1 and lie between 0 and 1, see Section 5.4.1. The code is as follows:

```r
multisite.marginalized <- nimbleCode({

  # -------------------------------------------------
  # Parameters:
  # phi111: survival-mov probability from state 11 to state 11
  # phi112: survival-mov probability from state 11 to state 12
  # phi121: survival-mov probability from state 12 to state 21
  # phi122: survival-mov probability from state 12 to state 22
  # phi211: survival-mov probability from state 21 to state 11
  # phi212: survival-mov probability from state 21 to state 12
  # phi221: survival-mov probability from state 22 to state 21
  # phi222: survival-mov probability from state 22 to state 22
  # det1: detection probability site 1
  # det2: detection probability site 2
  # pi11: init stat prob 11
  # pi12: init stat prob 12
```

```
# pi21: init stat prob 21
# pi22: init stat prob 22
# ---------------------------------------------------------
# States (S):
# 1 alive 11
# 2 alive 12
# 3 alive 21
# 4 alive 22
# 5 dead
# Observations (O):
# 1 not seen
# 2 seen at site 1
# 3 seen at site 2
# ---------------------------------------------------------

# priors
det1 ~ dunif(0, 1)
det2 ~ dunif(0, 1)
phi11[1:3] ~ ddirch(alpha[1:3]) # phi111, phi112, 1-sum
phi12[1:3] ~ ddirch(alpha[1:3]) # phi121, phi122, 1-sum
phi21[1:3] ~ ddirch(alpha[1:3]) # phi211, phi212, 1-sum
phi22[1:3] ~ ddirch(alpha[1:3]) # phi221, phi222, 1-sum

# probabilities of state z(t+1) given z(t)
gamma[1,1] <- phi11[1]
gamma[1,2] <- phi11[2]
gamma[1,3] <- 0
gamma[1,4] <- 0
gamma[1,5] <- phi11[3]
gamma[2,1] <- 0
gamma[2,2] <- 0
gamma[2,3] <- phi12[1]
gamma[2,4] <- phi12[2]
gamma[2,5] <- phi12[3]
gamma[3,1] <- phi21[1]
gamma[3,2] <- phi21[2]
gamma[3,3] <- 0
gamma[3,4] <- 0
```

```
gamma[3,5] <- phi21[3]
gamma[4,1] <- 0
gamma[4,2] <- 0
gamma[4,3] <- phi22[1]
gamma[4,4] <- phi22[2]
gamma[4,5] <- phi22[3]
gamma[5,1] <- 0
gamma[5,2] <- 0
gamma[5,3] <- 0
gamma[5,4] <- 0
gamma[5,5] <- 1

# probabilities of y(t) given z(t)
omega[1,1] <- 1 - det1
omega[1,2] <- det1
omega[1,3] <- 0
omega[2,1] <- 1 - det2
omega[2,2] <- 0
omega[2,3] <- det2
omega[3,1] <- 1 - det1
omega[3,2] <- det1
omega[3,3] <- 0
omega[4,1] <- 1 - det2
omega[4,2] <- 0
omega[4,3] <- det2
omega[5,1] <- 1
omega[5,2] <- 0
omega[5,3] <- 0

# initial state probs
for(i in 1:N) {
  # first state propagation
  init[i, 1:5] <- gamma[ y[i, first[i] ] - 1, 1:5 ]
}

# likelihood
for (i in 1:N){
  y[i,(first[i]+1):K] ~ dHMM(init = init[i,1:5],# data first[i]+1
```

```
                              probObs = omega[1:5,1:3], # obs
                              probTrans = gamma[1:5,1:5], # trans
                              len = K - first[i], # nb of occasions
                              checkRowSums = 0) # do not check
                                                # whether elements
                                                # in a row sum
                                                # to 1
  }
})
```

7.3.3 Results and interpretation

The raw results are:

```
##               mean   sd 2.5%  50% 97.5% Rhat n.eff
## det1          0.45 0.01 0.43 0.45  0.47 1.01   597
## det2          0.41 0.01 0.39 0.41  0.42 1.02   613
## phi11[1]      0.50 0.01 0.49 0.50  0.52 1.01   889
## phi11[2]      0.16 0.01 0.15 0.16  0.17 1.01  1103
## phi11[3]      0.34 0.01 0.32 0.34  0.35 1.00  1206
## phi12[1]      0.10 0.00 0.09 0.10  0.11 1.01   931
## phi12[2]      0.57 0.01 0.55 0.57  0.59 1.02   838
## phi12[3]      0.34 0.01 0.32 0.34  0.35 1.01  1007
## phi21[1]      0.37 0.03 0.31 0.36  0.42 1.00  1784
## phi21[2]      0.32 0.03 0.27 0.32  0.38 1.00  2196
## phi21[3]      0.31 0.03 0.25 0.31  0.38 1.00  1634
## phi22[1]      0.06 0.00 0.05 0.06  0.07 1.00  1424
## phi22[2]      0.63 0.01 0.61 0.63  0.64 1.00  2356
## phi22[3]      0.31 0.01 0.30 0.31  0.33 1.00  2067
```

which we can rearrange as in Table 7.1 to the results obtained by Pradel [2005] in his Table 1.

In Table 7.1, the first column gives the Transition made in t to $t + 1$ where M is for mid–Atlantic and C for Chesapeake. The second column gives the Condition that is whether the location at $t + 1$ is Equal or Not equal to the location at $t - 1$. The Pradel [2005] time-averaged

TABLE 7.1: Second-order (memory) estimates of transition probabilities.

Transition	Condition	Pradel (avg 1985–88)	NIMBLE
MM	Equal to t-1	0.57	0.50 (0.49–0.52)
MM	Not equal to t-1	0.33	0.36 (0.31–0.42)
MC	Equal to t-1	0.27	0.33 (0.27–0.38)
MC	Not equal to t-1	0.09	0.16 (0.15–0.17)
CM	Equal to t-1	0.21	0.10 (0.09–0.11)
CM	Not equal to t-1	0.05	0.06 (0.05–0.07)
CC	Equal to t-1	0.63	0.63 (0.61–0.64)
CC	Not equal to t-1	0.48	0.57 (0.55–0.59)

estimates are given in the third column. Our NIMBLE estimates (mean and 95% credible interval) are given in the fourth column.

In both analyses, memory is real: for each transition, the probability is higher when the destination at $t + 1$ matches where the bird was at $t - 1$ ("Equal to $t - 1$" rows) than when it doesn't ("Not equal to $t - 1$" rows). That's site fidelity or directional return. Our fit mirrors Pradel's fit especially well for staying in Chesapeake (CC, equal: 0.63 in both) and still shows a clear memory signal for staying in mid-Atlantic (MM, 0.50 vs 0.57 when equal). Moves also carry memory: MC is more likely when the bird was in C two steps back (0.33 vs 0.16), and CM shows a weaker but similar pattern (0.10 vs 0.06).

Where we differ is in the directional balance. Relative to Pradel's averages, our model leans more toward Chesapeake: we estimate higher MC probabilities (both equal and not equal; e.g., 0.33 vs 0.27 and 0.16 vs 0.09) and lower CM (equal) (0.10 vs 0.21). We also find a higher chance of staying in C (CC, not equal: 0.57 vs 0.48). Several of these differences are well outside our 95% credible intervals (e.g., CM equal 0.09–0.11 vs 0.21, MC not equal 0.15–0.17 vs 0.09). These discrepancies are likely explained by the difference in model structures: ours assumes constant parameters, whereas Pradel's allows them to vary over time. The big picture remains the same, though: movements are second–order and that memory is asymmetric, being strongest for persistence in Chesapeake and for moves toward it.

7.4 Accommodating individual heterogeneity

7.4.1 Motivation

I've worked with large carnivores for almost two decades, and they're what pulled me into HMMs – especially gray wolves. Wolves are social, live in hierarchical packs, and that structure shows up in demography. It also shows up in how we detect them: dominant individuals use trails and roads more often, and that's pretty much where we search for scats for DNA-based identification. The result is individual heterogeneity in detection, with some wolves being more detectable (via DNA) than others.

Shirley Pledger in a series of papers developed so-called mixture models in which individuals are assigned in two or more classes with class-specific survival/detection probabilities. If we ignore that variation, estimates can be biased and lack-of-fit tests light up. In Cubaynes et al. [2010], we used an HMM formulation to capture heterogeneity in detection [see also Pradel, 2009].

In what follows, I'll show how to build and fit these finite–mixture HMMs using simulated data so we control the truth. This gives a practical template for dealing with heterogeneity whenever behavior or status makes some individuals easier (or more difficult) to find than others.

7.4.2 Model and NIMBLE implementation

To build up our model, we first define the states and observations:

- States
    - alive in class 1 (A1)
    - alive in class 2 (A2)
    - dead (D)
- Observations

- not detected (1)
- detected (2)

Now we turn to writing our model. We start with the vector of initial states. Individuals enter as a mixture of A1 and A2:

$$\delta = \begin{pmatrix} \overset{z_t = A1}{\pi} & \overset{z_t = A2}{1-\pi} & \overset{z_t = D}{0} \end{pmatrix}$$

where π is the probability of being alive in A1, and $1-\pi$ is the probability of being in A2.

We proceed with the transition matrix that contains the survival probabilities:

$$\Gamma = \begin{pmatrix} \overset{z_t = A1}{\phi} & \overset{z_t = A2}{0} & \overset{z_t = D}{1-\phi} \\ 0 & \phi & 1-\phi \\ 0 & 0 & 1 \end{pmatrix} \begin{matrix} z_{t-1} = A1 \\ z_{t-1} = A2 \\ z_{t-1} = D \end{matrix}$$

where ϕ is the survival probability. In NIMBLE, we write:

```
# Transition matrix
gamma[1,1] <- phi        # A1(t)->A1(t+1)
gamma[1,2] <- 0          # A1(t)->A2(t+1)
gamma[1,3] <- 1 - phi    # A1(t)->D(t+1)
gamma[2,1] <- 0          # A2(t)->A1(t+1)
gamma[2,2] <- phi        # A2(t)->A2(t+1)
gamma[2,3] <- 1 - phi    # A2(t)->D(t+1)
gamma[3,1] <- 0          # D(t)->A1(t+1)
gamma[3,2] <- 0          # D(t)->A2(t+1)
gamma[3,3] <- 1          # D(t)->D(t+1)
```

Survival could be made heterogeneous; to do so, we'd need to amend the transition matrix as follows:

$$\Gamma = \begin{pmatrix} \overset{z_t = A1}{\phi(1 - \psi^{12})} & \overset{z_t = A2}{\phi\psi^{12}} & \overset{z_t = D}{1 - \phi} \\ \phi\psi^{21} & \phi(1 - \psi^{21}) & 1 - \phi \\ 0 & 0 & 1 \end{pmatrix} \begin{matrix} z_{t-1} = A1 \\ z_{t-1} = A2 \\ z_{t-1} = D \end{matrix}$$

where the ψ's are the probabilities for an individual to change class of heterogeneity, with ψ^{12} from A1 to A2, and ψ^{21} from A2 to A1.

Last step is the observation matrix, which we write as follows:

$$\Omega = \begin{pmatrix} \overset{y_t = 1}{1 - p^1} & \overset{y_t = 2}{p^1} \\ 1 - p^2 & p^2 \\ 1 & 0 \end{pmatrix} \begin{matrix} z_t = A1 \\ z_t = A2 \\ z_t = D \end{matrix}$$

where p^1 is detection for individuals in A1, and p^2 that of individuals in A2. In NIMBLE, this translates into:

```
# Observation matrix
omega[1,1] <- 1 - p1 # A1(t)->1(t) (non-detected)
omega[1,2] <- p1 # A1(t)->2(t) (detected)
omega[2,1] <- 1 - p2 # A2(t)->1(t)
omega[2,2] <- p2 # A2(t)->2(t)
omega[3,1] <- 1 # D(t)->1(t)
omega[3,2] <- 0 # D(t)->2(t)
```

At first detection, the observation "not detected" is impossible by construction (we condition on the first detection). We therefore use an initial observation matrix Ω_{init} with $p_1 = p_2$ set to 1 (not shown).

Instead of real data, we will use simulated data, which is useful to identify problems in our analysis as we will illustrate below. We simulate a finite–mixture scenario where individuals differ in detection but share the same survival. Each animal is assigned to one of two latent classes: a small, highly detectable group (class A1; proportion $\pi = 0.2$, $p_1 = 0.8$) and a larger, less detectable group (class A2; proportion $1 - \pi = 0.8$, $p_2 = 0.3$). Everyone starts captured at the first occasion (this mimics the CJS conditioning on first capture), then the latent alive state $z_{i,t}$ evolves with constant survival $\phi = 0.7$. If an individual i is alive at time t, it's observed with its own detection probability p_i (either p_1 or p_2 depending on class). The result is a matrix of 0/1 encounter histories y with built-in individual heterogeneity in detection, plus bookkeeping objects: `which_mixture` (true class), `detection` (each animal's p_i), and `z` (true alive states). We simulate the fate of 400 animals over 10 years.

```r
# Simulate encounter histories w ind het (finite mixture)

set.seed(1234)              # for reproducibility

# Parameters
phi <- 0.7                  # apparent survival (same for all)
prop_class1 <- 0.2          # mixture proportion pi: Pr(class 1)
p_class1 <- 0.8             # detection if in class 1 (high)
p_class2 <- 0.3             # detection if in class 2 (low)

nind <- 400                 # number of individuals
nyear <- 10                 # number of occasions

# Storage
z <- matrix(NA, nind, nyear)      # latent alive state
x <- matrix(NA, nind, nyear)      # observed detections
y <- matrix(NA, nind, nyear)      # final encounter histories (0/1)
first <- rep(1, nind)             # first-capture occasion
detection <- rep(NA, nind)        # each individual's p_i
which_mixture <- rep(NA, nind)    # latent class: 1 or 0 (class 2)

# Assign ind to classes, then give them class-specific detection
for (i in 1:nind) {
  which_mixture[i] <- rbinom(1, 1, prop_class1) # 1 w prob pi else 0
  detection[i] <- if (which_mixture[i] == 1) p_class1 else p_class2
}

# Generate latent states and observations
# Everyone is alive and detected at first capture (CJS conditioning)
for (i in 1:nind) {
  z[i, first[i]] <- 1
```

```r
  x[i, first[i]] <- 1

  # From the second occasion on:
  for (j in (first[i] + 1):nyear) {
    # Alive state: survive from previous if alive; else remain 0
    z[i, j] <- rbinom(1, 1, phi * z[i, j - 1])

    # Observation: if alive, detect with p_i; if dead, 0
    x[i, j] <- rbinom(1, 1, z[i, j] * detection[i])
  }
}

# Final encounter histories: replace NAs (before first) with 0
y <- x
y[is.na(y)] <- 0
```

The simulated data look like:

```r
head(y)
##      [,1] [,2] [,3] [,4] [,5] [,6] [,7] [,8] [,9] [,10]
## [1,]    1    0    0    0    0    0    0    0    0     0
## [2,]    1    0    0    0    0    0    0    0    0     0
## [3,]    1    0    0    0    0    0    0    0    0     0
## [4,]    1    0    1    0    0    0    0    0    0     0
## [5,]    1    0    0    0    0    0    0    0    0     0
## [6,]    1    1    0    0    0    0    0    0    0     0
```

Now when I fit the model above to these data, I get the following results:

```r
##       mean   sd 2.5%  50% 97.5% Rhat n.eff
## phi  0.70 0.01 0.67 0.70  0.72 1.01  1003
## pi   0.70 0.03 0.63 0.70  0.76 1.01   631
## pp1  0.45 0.02 0.40 0.45  0.50 1.02   431
## pp2  0.48 0.03 0.43 0.48  0.55 1.01   638
```

These estimates diverge markedly from the values used to simulate the data, see Table 7.2.

TABLE 7.2: Comparison of posterior estimates from NIMBLE with the data-generating values for a finite–mixture HMM.

Parameter	True value	Posterior mean (95% credible interval)
phi	0.7	0.70 (0.67–0.73)
pi	0.2	0.70 (0.63–0.76)
pp1	0.8	0.45 (0.40–0.49)
pp2	0.3	0.49 (0.43–0.55)

Why is that? The first issue is that the classes were permuted. In our model, nothing tells NIMBLE that the highly detectable individuals must belong to class A1 and the less detectable ones to class A2. The labeling of the classes is arbitrary. The interpretation only comes afterwards, by inspecting the parameter estimates. Now if we re-calculate π as the proportion of individuals in A1 as follows:

```r
samples <- rbind(mcmc.phipmix[[1]], mcmc.phipmix[[2]])
pi <- 1 - samples[,'pi']
mean(pi)
## [1] 0.3041
quantile(pi, probs = c(2.5, 97.5)/100)
##    2.5%   97.5%
## 0.2417 0.3727
```

We can get a sorted new Table 7.3.

TABLE 7.3: Comparison of posterior estimates from NIMBLE with the data-generating values for a finite–mixture HMM.

Parameter	True value	Posterior mean (95% credible interval)
phi	0.7	0.70 (0.67–0.73)
pi	0.2	0.30 (0.24–0.37)
pp1	0.8	0.49 (0.43–0.55)
pp2	0.3	0.45 (0.40–0.49)

Still, the detection estimates are off. There's a deeper issue here. The HMM formulation we used for capturing heterogeneity creates a limitation: individuals are not allowed to switch between classes over time (e.g., from A1 to A2), because the transition matrix does not permit it. In other words, any individual seen > 1 time (it's detected after the first observation occasion) can *never* change class assignments away from their initial value class assignment. Why this formulation fails? The problem is subtle but fundamental: in a HMM, transitions between states are governed by a Markov process. If the transition matrix says an individual in A1 must stay in A1 (or die), then the individual is permanently locked in that class. As a result, once an individual is assigned to a class through its initial latent state (e.g., A1), it can never change class during sampling. Even worse, any individual seen on multiple occasions (say, years 1 and 4) must stay alive during those years, the latent state cannot "jump" between classes without violating the transition constraints, therefore, the initial class assignment is fixed forever, and the MCMC sampler cannot explore the alternative class, no matter how much data supports it.

I must confess, I got stuck in that trap, and it was only because I used simulations that I could identify the problem. I then asked the NIMBLE team for help, and Daniel Turek came up with the explanation – thanks, Daniel!

Fortunately, there are several solutions to this problem:

- Marginalize over the latent states: Instead of sampling the latent states z, you can marginalize them out, which avoids the class-switching problem entirely. This is the principle behind the `nimbleEcology` package, see Section 3.8.3.2.

- Use a class-level latent variable: Rather than coding classes as HMM states, define a latent class assignment (e.g., `g[i] ~ dcat(pi[1:2])`) for each individual. The HMM then conditions on that fixed class, allowing for proper inference and switching during MCMC.

- Custom block sampler: Design a custom MCMC sampler that updates the entire latent history of an individual (`z[i, 1:K]`) at once. This allows for class-switching in a consistent way, though it's more advanced and computationally heavier.

Here I will adopt the first solution and use some material we saw in Section 3.8. In NIMBLE, we start by writing our own functions for the HMM likelihood, using the backward algorithm:

```r
# Integrate out latent states using custom density (forward algo)
# speed things up by pooling identical histories w pooled lik


# Get rid of individuals for which first==K
# individuals that are not first encountered at last occasion
mask <- which(first!=ncol(y))
y <- y[mask, ] # keep only these
first <- first[mask]
# if first == K (first seen at last occasion),
# no post-capture interval to inform phi/p;
# removing them simplifies forward recursion and
# avoids degenerate len=1 cases


# Pool encounter histories
y_weighted <- y %>%
  as_tibble() %>%
  group_by_all() %>% # group by entire history (all columns)
  summarise(size = n()) %>% # count how many inds share that history
  relocate(size) %>% # move "size" column to front
  as.matrix()
head(y_weighted)
size <- y_weighted[,1] # nb of inds w/ a particular history
y <- y_weighted[,-1] # pooled data: one row per unique history


# Assemble data and constants
# +1 because custom distribution expects categories 1/2 (not 0/1)
my.data <- list(y = y + 1)
my.constants <- list(N = nrow(y),
                     K = ncol(y),
                     first = first,
                     size = size,
                     one = 1)
# "one" is a dummy constant used in a dconstraint


# NIMBLE functions
```

```r
# Custom HMM *density* for pooled histories via forward algorithm
dwolfHMM <- nimbleFunction(
  run = function(x = double(1),
                 probInit = double(1), # initial states (delta)
                 probObs = double(2),  # obs (omega)
                 probObse = double(2), # obs used at first occasion
                 probTrans = double(2),# trans (gamma)
                 size = double(0),     # cardinal
                 len = double(0, default = 0), # nb sampling occ
                 log = integer(0, default = 0)) {
    # Initial forward probs:
    # multiply initial state probs by obs prob at first date x[1]
    # (probObs at t=first is 1 due to CJS conditioning;
    # and handled via 'probObse').
    alpha <- probInit[1:3] * probObse[1:3,x[1]]
    # * probObs[1:3,x[1]] == 1 due to conditioning on first detection
    for (t in 2:len) {
      # standard forward step:
      # alpha_t = (alpha_{t-1} * Gamma) .* Omega(:, y_t)
      alpha[1:3] <- (alpha[1:3] %*% probTrans[1:3,1:3]) *
        probObs[1:3,x[t]]
    }
    # Pooling: multiply the log-likelihood by 'size'
    logL <- size * log(sum(alpha[1:3]))
    returnType(double(0))
    if (log) return(logL)
    return(exp(logL))
  }
)


# Matching *random generator*
# (required by NIMBLE for custom distributions)
# Generates one synthetic history consistent with the HMM;
# not used in MCMC, but needed to register the dist.
rwolfHMM <- nimbleFunction(
  run = function(n = integer(),
                 probInit = double(1),
                 probObs = double(2),
```

```
                    probObse = double(2),
                    probTrans = double(2),
                    size = double(0),
                    len = double(0, default = 0)) {
  returnType(double(1))
  z <- numeric(len) # latent states
  # all individuals alive at t = 0 (by construction)
  z[1] <- rcat(n = 1, prob = probInit[1:3])
  y <- z
  y[1] <- 2 # all inds detected at t = 0 (CJS conditioning)
  for (t in 2:len){
    # state at t given state at t-1
    z[t] <- rcat(n = 1, prob = probTrans[z[t-1],1:3])
    # observation at t given state at t
    y[t] <- rcat(n = 1, prob = probObs[z[t],1:2])
  }
  return(y)
})


# Register the custom density/generator in the global environment
assign('dwolfHMM', dwolfHMM, .GlobalEnv)
assign('rwolfHMM', rwolfHMM, .GlobalEnv)
```

Now the NIMBLE code for the model, at last:

```
# Model code
hmm.phipmix <- nimbleCode({

  # ----------------------------------------------------------
  # Parameters:
  # pi: initial state probability A1    (mixture weight for class 1)
  # phi: survival probability           (shared across classes)
  # pp1: recapture probability A1        (detection for class 1)
  # pp2: recapture probability A2        (detection for class 2)
  # ----------------------------------------------------------
  # States (S):
  # 1 alive (A1)   -> class 1
  # 2 alive (A2)   -> class 2
```

```
# 3 dead (D)
# Observations (O):
# 1 neither seen nor recovered (0)
# 2 seen alive (1)
# ----------------------------------------------------------

# priors
phi ~ dunif(0, 1) # prior survival
# to avoid label switching (enforces an ordering)
one ~ dconstraint(pp1 < pp2)
pp1 ~ dunif(0, 1) # prior detection (class 1)
pp2 ~ dunif(0, 1) # prior detection (class 2)
pi ~ dunif(0, 1)  # prob init state 1 (mixture prop class 1)

# transition matrix (classes are persistent;
# no switching between A1 and A2)
gamma[1,1] <- phi       # A1(t)->A1(t+1)
gamma[1,2] <- 0         # A1(t)->A2(t+1)
gamma[1,3] <- 1 - phi   # A1(t)->D(t+1)
gamma[2,1] <- 0         # A2(t)->A1(t+1)
gamma[2,2] <- phi       # A2(t)->A2(t+1)
gamma[2,3] <- 1 - phi   # A2(t)->D(t+1)
gamma[3,1] <- 0         # D(t)->A1(t+1)
gamma[3,2] <- 0         # D(t)->A2(t+1)
gamma[3,3] <- 1         # D(t)->D(t+1)

# vector of initial state probs (mixture at first capture)
delta[1] <- pi          # A1(first)
delta[2] <- 1 - pi      # A2(first)
delta[3] <- 0           # D(first)

# observation matrix (det differs by class; dead never seen)
omega[1,1] <- 1 - pp1   # A1(t)->0(t)
omega[1,2] <- pp1       # A1(t)->1(t)
omega[2,1] <- 1 - pp2   # A2(t)->0(t)
omega[2,2] <- pp2       # A2(t)->1(t)
omega[3,1] <- 1         # D(t)->0(t)
omega[3,2] <- 0         # D(t)->1(t)
```

```r
  # 'omegae' is used only at first date inside custom density
  # to implement CJS cond (everyone detected at first capture)
  omegae[1,1] <- 0          # A1(t)->0(t)
  omegae[1,2] <- 1          # A1(t)->1(t)
  omegae[2,1] <- 0          # A2(t)->0(t)
  omegae[2,2] <- 1          # A2(t)->1(t)
  omegae[3,1] <- 1          # D(t)->0(t)
  omegae[3,2] <- 0          # D(t)->1(t)


  # likelihood
  for(i in 1:N) {
    # One weighted lik contrib per *unique* history,
    # using the forward algorithm implemented in dwolfHMM.
    y[i,first[i]:K] ~ dwolfHMM(probInit = delta[1:3],# initial states
                               probObs = omega[1:3,1:2], # obs
                               probObse=omegae[1:3,1:2],# obs first
                               probTrans = gamma[1:3,1:3], # trans
                               size = size[i], # weight
                               len = K - first[i] + 1) # nb occ
  }
})


# Initial values
# (cool thing, we do not need inits for latent states anymore!!)
initial.values <- function() list(phi = runif(1,0,1),
                                   pp1 = 0.3,
                                   pp2 = 0.8,
                                   pi = runif(1,0,1))
# 'one' is provided in constants; no init needed
# The pp1<pp2 constraint prevents label switching
# by fixing the class ordering


# Parameters to be monitored
parameters.to.save <- c("phi", "pp1", "pp2", "pi")


# Run NIMBLE
out <- nimbleMCMC(code = hmm.phipmix,
                  constants = my.constants,
```

```
            data = my.data,
            inits = initial.values,
            monitors = parameters.to.save,
            niter = n.iter,
            nburnin = n.burnin,
            nchains = n.chains)
```

7.4.3   Results and interpretation

The results in Table 7.4 look fine now.

In summary, when modeling unobservable individual heterogeneity (e.g., detection classes) in an HMM, avoid encoding class identity as a dynamic state. Instead, treat it as a fixed latent variable, or marginalize it out. Otherwise, the model is unable to explore the full posterior and will yield biased or unreliable estimates.

A question that remains is the number of classes we should use. In other words, why 2 classes and not 3 or 4? One option is to fit models with more classes and select among them [e.g., Cubaynes et al., 2012]. Alternatively, you can take a non-parametric route and let the data decide how many classes are needed; this is relatively easy in NIMBLE [see Turek et al., 2021].

TABLE 7.4: Posterior estimates vs. data-generating values for a finite-mixture HMM, when a marginalized likelihood is used.

Parameter	True value	Posterior mean (95% credible interval)
phi	0.7	0.72 (0.69–0.75)
pi	0.2	0.27 (0.10–0.46)
pp1	0.8	0.79 (0.65–0.93)
pp2	0.3	0.27 (0.17–0.35)

For broader context on individual heterogeneity, see Gimenez et al. [2018a]. Finally, remember that these hidden classes were introduced by Pledger and colleagues primarily to correct bias in survival or abundance when heterogeneity is ignored; interpreting the classes biologically after fitting should be done with caution.

8

Quantifying life history traits

8.1 Introduction

Capture-recapture models allow the estimation of key demographic parameters such as survival, recruitment, and dispersal. These parameters are central to life history theory, which seeks to explain how organisms allocate limited resources across survival, growth, and reproduction over their lifetime. Life-history traits, such as age at first breeding, breeding frequency, or lifespan, reflect trade-offs shaped by natural selection and environmental constraints.

Capture-recapture data are particularly suited to quantifying such traits in the wild, providing direct estimates that can be compared to theoretical predictions. As introduced in Section 5.5, states and transitions between them – whether disease, developmental, or breeding states – can be incorporated naturally into the HMM framework.

In this chapter, we focus on three examples: cause- and age-specific mortalities, age-specific breeding probabilities, and stopover duration. Together, they illustrate how capture-recapture HMMs can shed light on life-history strategies.

8.2 Assessing age- and cause-specific mortality

8.2.1 Motivation

Cause- and age-specific mortalities are key life-history traits, shaping population dynamics and informing management or conservation strategies. Yet in the wild, direct assignment of causes of death is

DOI: 10.1201/9781003244097-8

"

challenging as field observations are often incomplete, and carcass recovery is biased toward certain causes of death or habitats. If we ignore this uncertainty, estimates of mortality rates by cause or age class can be biased, and our understanding of population processes compromised.

The fix is to treat cause of death as a latent state. Following Koons et al. [2014], we model survival and transitions to cause-specific death states jointly within a HMM capture-recapture framework. To do so, we combine data on recaptures of live and recoveries of dead animals. Individuals can either survive to the next occasion or die from one of several competing causes, with probabilities that may depend on age.

Here, we use the case study on roe deer (*Capreolus capreolus*) in Koons et al. [2014]. The authors show that failing to account for uncertain cause assignment underestimated mortality from hunting and overestimated natural mortality, particularly in younger age classes.

Here, we define cause-specific death states, let survival and cause-specific mortality depend on age, and keep the rest of the HMM machinery unchanged in NIMBLE.

8.2.2 Model and NIMBLE implementation

For our example, we will use data on roe deer from a population in the enclosed forest of the Territoire d'Etude et d'Expérimentation of Trois Fontaines, in eastern France. Here, we focus on 556 known-age females (marked as fawns), of which 217 were recaptured at least once and 41 were deadly injured during handling, victim of car collisions, or recovered and reported by hunters (human-related mortalities). These data were kindly provided by Marlène Gamelon.

To build up our model, we first define the states and observations:

- States
  - A for alive
  - H for an individual that just died from human causes
  - NH for an individual that just died from a natural (non-human) cause
  - D for an individual that is already dead

- Observations
  - 1 for not seen
  - 2 for captured for the first time or recaptured
  - 3 for killed by human activities and reported

Now we turn to writing our model. We start with the vector of initial states. Individuals enter alive the study with probability 1:

$$\delta = \begin{pmatrix} \overset{z_t = A}{1} & \overset{z_t = H}{0} & \overset{z_t = NH}{0} & \overset{z_t = D}{0} \end{pmatrix}$$

We proceed with the transition matrix:

$$\Gamma = \begin{pmatrix} 1 - \mu_H - \mu_{NH} & \mu_H & \mu_{NH} & 0 \\ 0 & 0 & 0 & 1 \\ 0 & 0 & 0 & 1 \\ 0 & 0 & 0 & 1 \end{pmatrix} \begin{matrix} z_{t-1} = A \\ z_{t-1} = H \\ z_{t-1} = NH \\ z_{t-1} = D \end{matrix}$$

with column headers $z_t = A \quad z_t = H \quad z_t = NH \quad z_t = D$

where the transition probability from A at t to H at $t + 1$ is the human-related mortality probability μ_H and the transition probability from A at t to NH at $t + 1$ is the natural mortality probability μ_{NH}. Because an individual could not return to state A once dead, we fix these transitions to 0.

Following the paper, adult survival is age-specific: μ_H is different for individuals in their first year of life after birth ([0,1[) vs. age 1 onwards, while μ_{NH} has two different parameters for ages [0,1[and [1,2[then a linear trend with age from age 2 onward. We use a multinomial logit link function to ensure that mortality probabilities are estimated between 0 and 1 and sum up to 1, see Section 5.4.2. In NIMBLE, we code:

```
# Transition matrix

# Age effects
etaH[i,t] <- (alpha_H_01 * age_lt1[i,t] + alpha_H_ge1 * age_ge1[i,t])

etaNH[i,t] <- (alpha_NH_01 * age_lt1[i,t] + # if age is [0,1[
               alpha_NH_12 * age_1to2[i,t] + # if age is [1,2[
```

```r
                # linear trend if age >= 2
                (beta0_NH+beta1_NH*age_std_mu2[i,t])*age_ge2[i,t])

# Transitions from A
den[i,t] <- 1 + exp(etaH[i,t]) + exp(etaNH[i,t])   # denominator
gamma[i,t,1,1] <- 1 / den[i,t]                      # stay alive
gamma[i,t,1,2] <- exp(etaH[i,t]) / den[i,t]         # muH
gamma[i,t,1,3] <- exp(etaNH[i,t]) / den[i,t]        # muNH
gamma[i,t,1,4] <- 0                                 # always 0

# From H, NH, D (always go to D)
gamma[i,t,2,1] <- 0
gamma[i,t,2,2] <- 0
gamma[i,t,2,3] <- 0
gamma[i,t,2,4] <- 1
gamma[i,t,3,1] <- 0
gamma[i,t,3,2] <- 0
gamma[i,t,3,3] <- 0
gamma[i,t,3,4] <- 1
gamma[i,t,4,1] <- 0
gamma[i,t,4,2] <- 0
gamma[i,t,4,3] <- 0
gamma[i,t,4,4] <- 1
```

In the code above, age is passed in the data and created as follows:

```r
# Build an N x K age matrix
age <- matrix(0, nrow = nrow(obs_matrix), ncol = K)
for (i in 1:nrow(obs_matrix)) {
  for (t in first[i]:K) {
    age[i, t] <- t - first[i]   # age 0 at first capture
  }
}
# Define age [0,1[ and age >=1
for (i in 1:nrow(obs_matrix)) {
  for (t in first[i]:K) {
    age_val <- age[i, t]
```

```r
    if (age_val == 0) { # age < 1
      age_lt1[i, t] <- 1
      age_ge1[i, t] <- 0
    } else {            # age >= 1
      age_lt1[i, t] <- 0
      age_ge1[i, t] <- 1
    }
  }
}
# Define age >1 and age [1,2[
for (i in 1:nrow(obs_matrix)) {
  for (t in first[i]:K) {
    age_val <- age[i, t]
    age_ge2[i, t] <- as.integer(age_val > 1) # age > 1 or 2 onwards
    age_1to2[i, t] <- as.integer(age_val == 1) # age in [1,2[
  }
}
```

Last step is the observation matrix:

$$
\Omega = \begin{array}{ccc} y_t = 1 & y_t = 2 & y_t = 3 \end{array} \\
\Omega = \begin{pmatrix} 1-p & p & 0 \\ 1-\lambda & 0 & \lambda \\ 1 & 0 & 0 \\ 1 & 0 & 0 \end{pmatrix} \begin{array}{l} z_t = A \\ z_t = H \\ z_t = NH \\ z_t = D \end{array}
$$

where a live individual can be detected with probability p, or not with probability $1 - p$. We include time-dependence and age on p, with a different parameter for 1 year-old individuals vs. individuals older than 1 year of age. In NIMBLE, we write:

```r
pA[i,t] <- p_ageLE1[t] * age_lt1[i,t] + p_ageGT1[t] * age_ge1[i,t]
```

An individual that just died from human causes can be recovered and reported with probability λ, or not with probability $1 - \lambda$. Because tag recovery protocols were constant over the course of the study, the authors considered that parameter to be constant over time and across age classes.

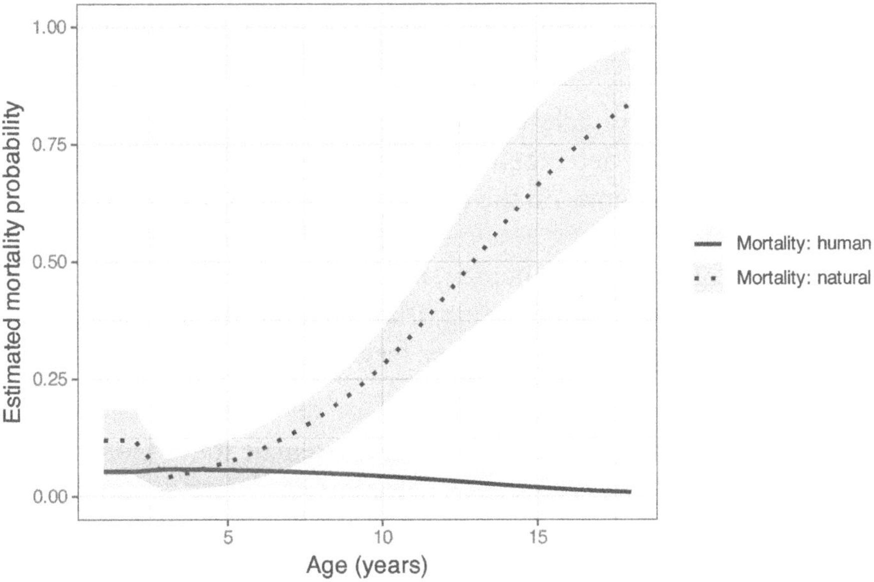

FIGURE 8.1: Age-specific mortality probabilities for human-related (solid line) and natural (dashed line) causes in female roe deer at Trois-Fontaines, France, from 1985 to 2013. Shaded polygons represent 95% credible intervals.

8.2.3 Results and interpretation

The results are given in Figure 8.1. We see senescence in natural mortality from age 2 onwards, as well as well as a slightly higher human-related mortality in the first year of life after birth. These patterns are similar to those obtained by Koons et al. [2014], check out their Figure 4.

8.3 Quantifying age-specific breeding

8.3.1 Motivation

Breeding is not a simple yes/no coin flip. Who breeds this year depends on what happened last year. In long-lived mammals, adult survival is high and fairly stable, but breeding probability can differ according to recent reproductive history and with how reliably we detect mothers and their offspring. If we blur these pieces and treat breeding status as perfectly known we risk biasing both survival and breeding estimates.

Couet et al. [2019] give a clear template with female bottlenose dolphins. They show that calf detectability differs by age and a female's breeding probability depends on the outcome of her previous calf. Rather than assuming states are observed without error, they use a HMM to keep true breeding state hidden, model misclassification/detection explicitly, and tie breeding this year to last year's outcome. That way, uncertainty about calf presence and age feeds correctly into demographic estimates.

Here, we follow that logic: we define hidden female states (nonbreeder vs. breeder with calf age classes), let detection vary by calf age, and make breeding probability state-dependent on the previous year's result.

8.3.2   Model and NIMBLE implementation

For our example, we'll work with data from a population of bottlenose dolphins (*Tursiops truncatus*) inhabiting the French waters of the Normano-Breton Gulf in the English Channel. We have 106 encounter histories between 2004 and 2016, made of photographs taken from a boat between July and November, which corresponds to the birth peak and the best weather conditions for monitoring. These data were shared by Pauline Couet.

To build up our model, we first define the states and observations:

- States
  - NB for a non-breeding adult female
  - Byoy for a breeding adult female with a young of the year
  - Bc1 for a breeding adult female with a 1-year-old calf
  - Bc3 for a breeding adult female with a 3-year-old calf
  - Bc2 for a breeding adult female with a 2-year-old calf
  - D for a dead female
- Observations
  - 1 for not seen

- 2 for seen alone
- 3 for seen with a young of the year
- 4 for seen with a calf (1 to 3 years old)

Now we turn to writing our model. We start with the vector of initial states:

$$\delta = \begin{array}{cccccc} NB & Byoy & Bc1 & Bc2 & Bc3 & D \\ \left(\pi_1 \right. & \pi_2 & \pi_3 & \pi_4 & \pi_5 & \left. 0 \right) \end{array}$$

where $\pi_5 = 1 - \sum_{j=1}^{4} \pi_j$.

We proceed with the transition matrix given by Couet et al. [2019]:

$$\Gamma = \begin{array}{cccccc} z_t = NB & z_t = Byoy & z_t = Bc1 & z_t = Bc2 & z_t = Bc3 & z_t = D \\ \phi_A(1-\beta_1) & \phi_A\beta_1 & 0 & 0 & 0 & 1-\phi_A \\ 0 & 0 & \phi_A\phi_Y & \phi_A(1-\phi_Y) & 0 & 1-\phi_A \\ \phi_A(1-\beta_1) & \phi_A\beta_1 & 0 & 0 & 0 & 1-\phi_A \\ \phi_A(1-\beta_2) & \phi_A\beta_2 & 0 & 0 & 0 & 1-\phi_A \\ \phi_A(1-\beta_2) & \phi_A\beta_2 & 0 & 0 & 0 & 1-\phi_A \\ 0 & 0 & 0 & 0 & 0 & 1 \end{array} \begin{array}{l} z_{t-1} = NB \\ z_{t-1} = Byoy \\ z_{t-1} = Bc1 \\ z_{t-1} = Bc2 \\ z_{t-1} = Bc3 \\ z_{t-1} = D \end{array}$$

Rather than attempting to explain the full complexity of this matrix at once, I will follow the authors' suggestion and break it down into four steps, each corresponding to a biological event in the life cycle of female bottlenose dolphins: (1) female survival Φ_A, (2) young survival Φ_Y, (3) young aging $AGING$, and (4) breeding $BREED$, each step being conditional on the previous ones. To do so, we introduce some intermediate states:

- Intermediate states
  - Byoy-D for a breeding adult female that had lost her young of the year
  - Bc1-D for a breeding adult female that had lost her 1-year-old calf
  - Bc2-D for a breeding adult female that had lost her 2-year-old calf
  - Bc3-leave Breeding adult female that raised her calf to the age of 3

The four corresponding matrices are given below, starting with the female survival matrix:

$$
\Phi_A =
\begin{array}{c}
\begin{array}{cccccc}
\text{NB} & \text{Byoy} & \text{Bc1} & \text{Bc2} & \text{Bc3} & \text{D}
\end{array} \\
\left(
\begin{array}{cccccc}
\phi_A & 0 & 0 & 0 & 0 & 1-\phi_A \\
0 & \phi_A & 0 & 0 & 0 & 1-\phi_A \\
0 & 0 & \phi_A & 0 & 0 & 1-\phi_A \\
0 & 0 & 0 & \phi_A & 0 & 1-\phi_A \\
0 & 0 & 0 & 0 & \phi_A & 1-\phi_A \\
0 & 0 & 0 & 0 & 0 & 1
\end{array}
\right)
\begin{array}{l}
\text{NB} \\ \text{Byoy} \\ \text{Bc1} \\ \text{Bc2} \\ \text{Bc3} \\ \text{D}
\end{array}
\end{array}
$$

An adult female survives from t to $t+1$ with probability ϕ_A or dies with $1-\phi_A$. We go on with the young survival matrix:

$$
\Phi_Y =
\begin{array}{c}
\begin{array}{ccccccccc}
\text{NB} & \text{Byoy} & \text{Byoy-D} & \text{Bc1} & \text{Bc1-D} & \text{Bc2} & \text{Bc2-D} & \text{Bc3-leave} & \text{D}
\end{array} \\
\left(
\begin{array}{ccccccccc}
1 & 0 & 0 & 0 & 0 & 0 & 0 & 0 & 0 \\
0 & \phi_Y & 1-\phi_Y & 0 & 0 & 0 & 0 & 0 & 0 \\
0 & 0 & 0 & \phi_Y & 1-\phi_Y & 0 & 0 & 0 & 0 \\
0 & 0 & 0 & 0 & 0 & \phi_Y & 1-\phi_Y & 0 & 0 \\
0 & 0 & 0 & 0 & 0 & 0 & 0 & 1 & 0 \\
0 & 0 & 0 & 0 & 0 & 0 & 0 & 0 & 1
\end{array}
\right)
\begin{array}{l}
\text{NB} \\ \text{Byoy} \\ \text{Bc1} \\ \text{Bc2} \\ \text{Bc3} \\ \text{D}
\end{array}
\end{array}
$$

Offspring survive with probability ϕ_Y or die with $1-\phi_Y$. If the calf survives, the female remains in the same breeding state. If it dies, she moves to an intermediate "dead-offspring" state (BYOY-D, Bc1-D, or Bc2-D) so that later steps can condition on that loss. For 3-year-old calves, mortality cannot be distinguished from emancipation (leaving the mother), so we add an intermediate Bc3-leave state and fix survival to 1 there (this parameter is otherwise unidentifiable). As in the paper, I distinguish young-of-the-year and 1-year-old calf survival from 2-year-old calf survival. We go on with the young aging matrix:

$$
AGING =
\begin{array}{c}
\begin{array}{ccccccccc}
\text{NB} & \text{Byoy} & \text{Byoy-D} & \text{Bc1} & \text{Bc1-D} & \text{Bc2} & \text{Bc2-D} & \text{Bc3-leave} & \text{D}
\end{array} \\
\left(
\begin{array}{ccccccccc}
1 & 0 & 0 & 0 & 0 & 0 & 0 & 0 & 0 \\
0 & 1 & 0 & 0 & 0 & 0 & 0 & 0 & 0 \\
0 & 0 & 1 & 0 & 0 & 0 & 0 & 0 & 0 \\
0 & 0 & 0 & 0 & 1 & 0 & 0 & 0 & 0 \\
0 & 0 & 0 & 0 & 1 & 0 & 0 & 0 & 0 \\
0 & 0 & 0 & 0 & 0 & 0 & 1 & 0 & 0 \\
0 & 0 & 0 & 0 & 0 & 0 & 1 & 0 & 0 \\
0 & 0 & 0 & 0 & 0 & 0 & 0 & 1 & 0 \\
0 & 0 & 0 & 0 & 0 & 0 & 0 & 0 & 1
\end{array}
\right)
\begin{array}{l}
\text{NB} \\ \text{Byoy} \\ \text{Byoy-D} \\ \text{Bc1} \\ \text{Bc1-D} \\ \text{Bc2} \\ \text{Bc2-D} \\ \text{Bc3-leave} \\ \text{D}
\end{array}
\end{array}
$$

The progression of calf age classes between t and $t+1$ is deterministic, all relevant age-advancing transitions are set to 1, and there are no parameters estimated in this matrix. Last, here is the breeding matrix:

$$BREED = \begin{pmatrix} 1-\beta_1 & \beta_1 & 0 & 0 & 0 & 0 \\ 0 & 0 & 1 & 0 & 0 & 0 \\ 0 & 0 & 0 & 1 & 0 & 0 \\ 0 & 0 & 0 & 0 & 1 & 0 \\ 1-\beta_1 & \beta_1 & 0 & 0 & 0 & 0 \\ 1-\beta_1 & \beta_1 & 0 & 0 & 0 & 0 \\ 1-\beta_2 & \beta_2 & 0 & 0 & 0 & 0 \\ 1-\beta_2 & \beta_2 & 0 & 0 & 0 & 0 \\ 0 & 0 & 0 & 0 & 0 & 1 \end{pmatrix} \begin{matrix} \text{NB} \\ \text{Byoy} \\ \text{Bc1} \\ \text{Bc2} \\ \text{Byoy-D} \\ \text{Bc1-D} \\ \text{Bc2-D} \\ \text{Bc3-leave} \\ \text{D} \end{matrix}$$

Breeding at $t+1$ occurs with probability β (otherwise $1-\beta$). Breeding can depend on the female's state, namely non-breeding females and females that had just lost their young of the year or 1-year-old calf (β_1) vs. females that had lost their 2-year-old calf or raised it until the age of 3 (β_2).

Overall, the transition matrix is the product of all four matrices. In NIMBLE, we could code directly the full matrix, but I take this opportunity to illustrate how to code the splitting in several ecological processes:

```
# Transition matrix

# PHIA is 6x6
for (i in 1:5) {
  PHIA[i, i] <- adultsurvival
  PHIA[i, 6] <- 1 - adultsurvival
}
PHIA[6, 6] <- 1

# PHIY is 6x9
# Row 1: NB stays NB
PHIY[1, 1] <- 1
```

```r
# Row 2: Byoy
PHIY[2, 2] <- youngsurvival[1] # YOY survives → becomes Bc1
PHIY[2, 3] <- 1 - youngsurvival[1] # YOY dies → becomes Byoy-D
# Row 3: Bc1
PHIY[3, 4] <- youngsurvival[1] # 1-year-old survives → Bc2
PHIY[3, 5] <- 1 - youngsurvival[1] # dies → Bc1-D
# Row 4: Bc2
PHIY[4, 6] <- youngsurvival[2] # 2-year-old survives → Bc3-L
PHIY[4, 7] <- 1 - youngsurvival[2] # dies → Bc2-D
# Row 5: Bc3 always emancipates
PHIY[5, 8] <- 1
# Row 6: dead female stays dead
PHIY[6, 9] <- 1


# AGING is 9x9
AGING[1, 1] <- 1    # NB → NB
AGING[2, 2] <- 1    # Byoy → Bc1
AGING[3, 5] <- 1    # Byoy-D → Byoy-D
AGING[4, 3] <- 1    # Bc1 → Bc2
AGING[5, 6] <- 1    # Bc1-D → Bc1-D
AGING[6, 4] <- 1    # Bc2 → Bc3
AGING[7, 7] <- 1    # Bc2-D → Bc2-D
AGING[8, 8] <- 1    # Bc3-L → Bc3-L
AGING[9, 9] <- 1    # D → D


# BREED is 9x6
# Row 1: NB → Byoy or NB
BREED[1, 2] <- beta[1]
BREED[1, 1] <- 1 - beta[1]
# Row 2: Bc1 → Bc1
BREED[2, 3] <- 1
# Row 3: Bc2 → Bc2
BREED[3, 4] <- 1
# Row 4: Bc3 → Bc3
BREED[4, 5] <- 1
# Row 5: Byoy-D → Byoy or NB
```

```
BREED[5, 2] <- beta[1]
BREED[5, 1] <- 1 - beta[1]
# Row 6: Bc1-D → Byoy or NB
BREED[6, 2] <- beta[1]
BREED[6, 1] <- 1 - beta[1]
# Row 7: Bc2-D → Byoy or NB
BREED[7, 2] <- beta[2]
BREED[7, 1] <- 1 - beta[2]
# Row 8: Bc3-L → Byoy or NB
BREED[8, 2] <- beta[2]
BREED[8, 1] <- 1 - beta[2]
# Row 9: D → D
BREED[9, 6] <- 1


gamma[1:6,1:6] <- PHIA[1:6,1:6] %*% PHIY[1:6,1:9] %*%
   AGING[1:9,1:9] %*% BREED[1:9,1:6]
```

Last step is the observation matrix that we can also split in two steps, given these intermediate observations:
- ND: for not detected
- DNB: for a non-breeding female detected at time t
- DByoy: for a breeding female with a young of the year detected at time t
- DBc: for a breeding female with a 1-year-old calf detected at time t
- DBc2: for a breeding female with a 2-year-old calf detected at time t
- DBc3: for a breeding female with a 3-year-old calf detected at time t

First, the detection step:

$$DETECTION = \begin{array}{c} \\ \text{NB} \\ \text{Byoy} \\ \text{Bc1} \\ \text{Bc2} \\ \text{Bc3} \\ \text{D} \end{array} \begin{pmatrix} 1-p_1 & p_1 & 0 & 0 & 0 & 0 \\ 1-p_2 & 0 & p_2 & 0 & 0 & 0 \\ 1-p_2 & 0 & 0 & p_2 & 0 & 0 \\ 1-p_1 & 0 & 0 & 0 & p_1 & 0 \\ 1-p_1 & 0 & 0 & 0 & p_1 & 0 \\ 1 & 0 & 0 & 0 & 0 & 0 \end{pmatrix}$$

with columns ND, DNB, DByoy, DBc, DBc2, DBc3.

which contains the probability p of detecting the female given her latent state at t: p_2 is for females with a young of the year or a 1-year-old calf and p_1 for non-breeding females or those with an older calf. Then the observation matrix:

$$OBS = \begin{array}{cccc} 1 & 2 & 3 & 4 \end{array}$$

$$OBS = \begin{pmatrix} 1 & 0 & 0 & 0 \\ 0 & 1 & 0 & 0 \\ 0 & 1-q_1 & q_1 & 0 \\ 0 & 1-q_2 & 0 & q_2 \\ 0 & 1-q_2 & 0 & q_2 \\ 0 & 1-q_2 & 0 & q_2 \end{pmatrix} \begin{array}{l} \text{ND} \\ \text{DNB} \\ \text{DByoy} \\ \text{DBc} \\ \text{DBc2} \\ \text{DBc3} \end{array}$$

where, conditional on being in a breeding state, there is a probability q that a calf is seen vs. not seen. Calf age is not identifiable from photographs, so observations are seen or not seen, but q may still vary by calf age (here young of the year vs. older). This is coded in NIMBLE as follows:

```
# Observation matrix


# Detection probabilities
DETECTION[1, 1] <- 1 - p[1]        # NB not detected
DETECTION[1, 2] <- p[1]            # NB detected
DETECTION[2, 1] <- 1 - p[2]
DETECTION[2, 3] <- p[2]            # Byoy detected
DETECTION[3, 1] <- 1 - p[2]
DETECTION[3, 4] <- p[2]            # Bc1 detected
DETECTION[4, 1] <- 1 - p[1]
DETECTION[4, 5] <- p[1]            # Bc2 detected
DETECTION[5, 1] <- 1 - p[1]
DETECTION[5, 6] <- p[1]            # Bc3 detected
DETECTION[6, 1] <- 1            # Dead never detected


# OBS is 6x4
# Not detected -> not sighted
OBS[1, 1] <- 1
```

```
# Detected NB → alone
OBS[2, 2] <- 1
# Detected Byoy → YOY visible or not
OBS[3, 2] <- 1 - q[1]      # alone
OBS[3, 3] <- q[1]          # with YOY
# Detected Bc1 → calf visible or not
OBS[4, 2] <- 1 - q[2]
OBS[4, 4] <- q[2]
# Detected Bc2
OBS[5, 2] <- 1 - q[2]
OBS[5, 4] <- q[2]
# Detected Bc3
OBS[6, 2] <- 1 - q[2]
OBS[6, 4] <- q[2]


omega[1:6,1:4] <- DETECTION[1:6,1:6] %*% OBS[1:6,1:4]
```

8.3.3   Results and interpretation

The raw results are:

```
##                     mean   sd 2.5%  50% 97.5% Rhat n.eff
## adultsurvival       0.96 0.01 0.94 0.96  0.98 1.00  2069
## beta[1]             0.35 0.04 0.28 0.35  0.45 1.03   656
## beta[2]             0.57 0.10 0.39 0.57  0.78 1.00  1009
## p[1]                0.63 0.03 0.58 0.63  0.69 1.00  1764
## p[2]                0.79 0.03 0.72 0.79  0.86 1.00  1123
## pi[1]               0.49 0.06 0.37 0.49  0.60 1.01  2743
## pi[2]               0.26 0.05 0.17 0.26  0.37 1.00  3412
## pi[3]               0.08 0.03 0.03 0.08  0.16 1.00  2225
## pi[4]               0.14 0.04 0.07 0.14  0.23 1.00  1931
## pi[5]               0.03 0.03 0.00 0.02  0.10 1.02  1315
## q[1]                0.59 0.05 0.49 0.59  0.69 1.01  1091
## q[2]                0.80 0.05 0.70 0.80  0.89 1.01   818
## youngsurvival[1]    0.63 0.05 0.55 0.63  0.73 1.00   808
## youngsurvival[2]    0.69 0.08 0.53 0.69  0.83 1.00  1702
```

TABLE 8.1: Age-specific estimates of breeding probabilities. Comparison of our NIMBLE estimates with Couet et al. (2019). "NR" denotes quantities not reported in the paper. "NB" is for non-breeding and "yoy" for young-of-the-year. Adult female detection on the paper side is a rough time-average read by eye from the authors' Figure 3 (2005–2016).

Parameter	NIMBLE	Couet
Adult female survival	0.96 (0.94–0.98)	0.97 (0.96–0.98)
Calf observation, yoy	0.59 (0.49–0.69)	0.58 (0.46–0.68)
Calf observation, 1–3y	0.80 (0.70–0.89)	0.79 (0.59–0.90)
Calf survival (yoy)	0.63 (0.55–0.73)	0.66 (0.50–0.78)
Calf survival (1y)	0.69 (0.53–0.83)	0.45 (0.29–0.61)
Female detection: NB or calf 2–3y	0.63 (0.58–0.69)	≈ 0.47
Female detection: yoy or 1-year calf	0.79 (0.72–0.86)	≈0.61
Initial state NB	0.49 (0.37–0.60)	NR
Initial state Byoy	0.26 (0.17–0.37)	NR
Initial state Bc1	0.08 (0.03–0.16)	NR
Initial state Bc2	0.14 (0.07–0.23)	NR
Initial state Bc3	0.03 (0.00–0.10)	NR

which we can re-arrange as in Table 8.1 to compare them to the estimates obtained by Couet et al. [2019].

Our NIMBLE estimates track the published values closely. Calf survival shows some discrepancies, but credible and confidence intervals overlap, so I would not worry too much. Note also that our model differs slightly: we held adult female detection constant in time, whereas the paper models it as time-dependent.

Offspring are often missed even when a female is breeding, especially young of the year, which are harder to see than older calves. Offspring detection is well below 1. Adult detection also depends on the female's state: females with a young of the year or 1-year calf are more detectable than non-breeders or those with older calves.

Adult female survival is high. Calf survival ranks young-of-the-year $\geq$ 1-year. Breeding probability depends on recent history: it's lower after non-breeding or losing a young-of-the-year/1-year calf, and higher after losing a 2-year calf or successfully raising a calf to age 3.

8.4 Estimating stopover duration

8.4.1 Motivation

Stopover is not just a logistical pause; it's a stage-specific timing trait in a migratory life cycle. Like age at first reproduction or length of a breeding attempt, the duration of stopover shapes exposure to predators, energy balance, and downstream reproductive timing. Two core questions in migration ecology are: (1) how long individuals stop over, and (2) whether groups differ in their stopover behavior. Treating stopover as a life-history trait puts those questions center stage.

The catch is that arrivals and departures are hidden and detection is imperfect. If we only look at when individuals are seen, we bias both the entry curve and the distribution of stay lengths. As reviewed in Guérin et al. [2017], a HMM fixes this by keeping states hidden and observations separate: for example, we may use three latent states: not yet arrived, arrived (present), departed; and let two transition probabilities drive dynamics: recruitment (from not yet arrived to arrived) and staying (from arrived to arrived). Detection operates only in state arrived. This so-called time-effect parameterization assumes staying depends on calendar date, not time since arrival, which is often the right first step.

With this setup we can estimate daily entry and staying probabilities, derive expected stopover duration for an arrival date, and test for group differences in behavior.

8.4.2 Model and NIMBLE implementation

We illustrate the model with marbled newts (*Triturus marmoratus*) monitored in 2013 at a permanent pond near Vigneux-de-Bretagne (Loire-Atlantique, western France). Intensive capture-recapture yielded 713 PIT-tagged newts (412 males, 301 females). Captures within the same week were pooled to create 12 equally spaced occasions (11 weekly samples plus a leading "empty" occasion), so all individuals start in the "not yet arrived" state. These data were shared by Sandra Guérin.

To build up our model, we first define the states and observations:

- States
  - NYA for not yet arrived in the pond
  - ARR for arrived in the pond
  - DEP departed from the pond

- Observations
  - 1 for not captured
  - 2 for captured

Now we turn to writing our model. We start with the vector of initial states. At the start of the study, everyone is in NYA, so the initial state vector is:

$$\delta = \begin{array}{ccc} NYA & ARR & DEP \end{array} \\ \left(\begin{array}{ccc} 1 & 0 & 0 \end{array} \right)$$

We proceed with the transition matrix, which is governed by two processes: recruitment r is the probability that an individual in NYA enters the pond and staying s is the probability that an individual in ARR remains in the pond. State DEP is absorbing. We have:

$$\Gamma = \begin{array}{cccc} z_t = NYA & z_t = ARR & z_t = DEP \\ \left(\begin{array}{ccc} 1-r & r & 0 \\ 0 & s & 1-s \\ 0 & 0 & 1 \end{array} \right) & \begin{array}{c} z_{t-1} = NYA \\ z_{t-1} = ARR \\ z_{t-1} = DEP \end{array} \end{array}$$

We make both parameters r and s sex and time-dependent, which we code in NIMBLE:

```
for (j in 1:(K-1)){
  # prior probability of arriving in the pond / 1 = female
  logit(r[1,j]) <- beta[1,1] + beta[1,2] * (j-4.5)/2.45
```

```
  # prior probability of arriving in the pond / 2 = male
  logit(r[2,j]) <- beta[2,1] + beta[2,2] * (j-4.5)/2.45
  # prior retention probability (staying) / 1 = female
  s[1,j] ~ dunif(0, 1)
  # prior retention probability (staying) / 2 = male
  s[2,j] ~ dunif(0, 1)
}
for (i in 1:N){
  for (j in 1:(K-1)){
    gamma[1,1,i,j] <- 1 - r[sex[i],j]
    gamma[1,2,i,j] <- r[sex[i],j]
    gamma[1,3,i,j] <- 0
    gamma[2,1,i,j] <- 0
    gamma[2,2,i,j] <- s[sex[i],j]
    gamma[2,3,i,j] <- 1 - s[sex[i],j]
    gamma[3,1,i,j] <- 0
    gamma[3,2,i,j] <- 0
    gamma[3,3,i,j] <- 1
  }
}
```

Graphically, this corresponds to:

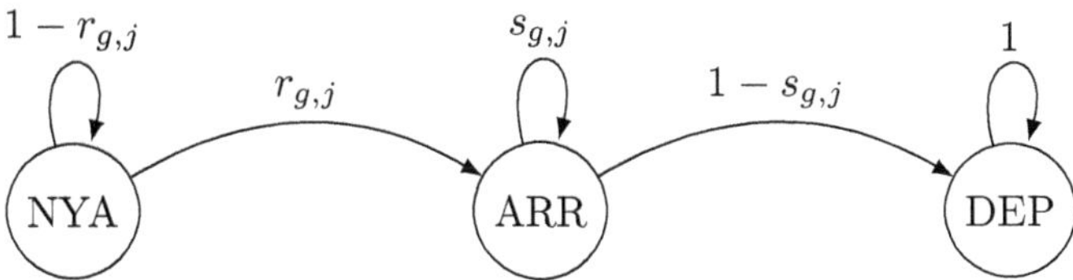

Last step is the observation matrix:

$$\Omega = \begin{array}{cc} y_t = 1 & y_t = 2 \end{array} \\ \begin{pmatrix} 1 & 0 \\ 1-p & p \\ 1 & 0 \end{pmatrix} \begin{array}{l} z_t = NYA \\ z_t = ARR \\ z_t = DEP \end{array}$$

where only individuals in state ARR are observable with probability p.

The likelihood and priors are handled as usual.

Using MCMC draws from the posterior distribution of the staying probability s, we can calculate the expected stopover duration $\mathbb{E}[D]$ which is the number of occasions present after arrival. With constant s, this is given by:

$$\mathbb{E}[D] \;=\; \frac{1}{1-s}.$$

Why is that? Because the length of stay (counting the arrival occasion) is geometric. After arriving, you are present this week with probability 1 (that's 1 week). You are still present next week with probability s. You are still present the week after with probability s^2. And so on. So the expected number of occasions present is the sum of those probabilities:

$$\mathbb{E}[D] \;=\; 1 + s + s^2 + \cdots \;=\; \frac{1}{1-s}.$$

When s varies with date, the "length of stay" after arrival is no longer geometric with a single parameter, it's a non-homogeneous geometric process with week-specific stay probabilities s_t. If an individual arrives on date τ, the chance it is still present k occasions later is the product of the stay probabilities along that run:

$$Pr(D > k \mid \text{arrive at } \tau) \;=\; \prod_{j=0}^{k-1} s_{\tau+j}.$$

Therefore, the expected number of occasions present (counting the arrival occasion) is the sum of those tail probabilities:

$$\mathbb{E}[D \mid \text{arrive at } \tau] \;=\; \sum_{k=0}^{\infty} \prod_{j=0}^{k-1} s_{\tau+j}.$$

with the empty product for $k = 0$ equal to 1, or in other words the sum of the probabilities of "surviving" each additional week in ARR under the calendar-date effects.

8.4.3   Results and interpretation

Our NIMBLE estimates match pretty closely those obtained by Guérin et al. [2017], compare our Figure 8.2 with the authors' Figure 2.

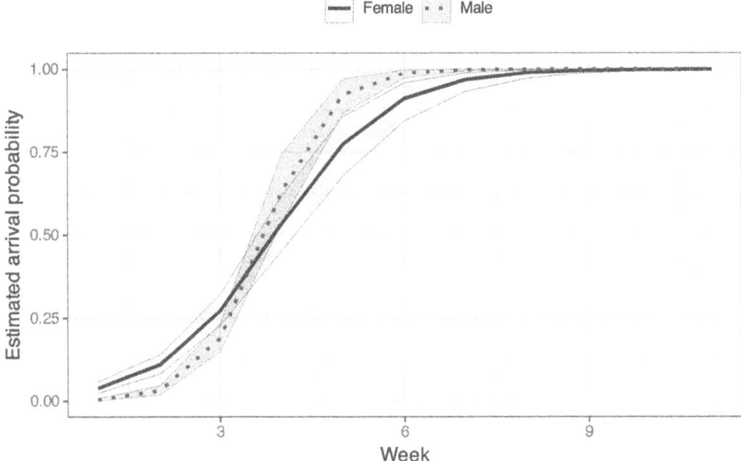

FIGURE 8.2: Arrival probability by sex and week. Means and 95% credible intervals (shaded areas) are provided.

Early in the season, females arrive first: during the first three capture occasions their arrival probability exceeds that of males (Figure 8.2). By occasion 7, essentially all individuals have entered the pond.

Females show longer stopovers than males for most of the season; near the end, the pattern reverses (Figure 8.3). Individuals arriving early in the breeding season stay longer than those arriving later.

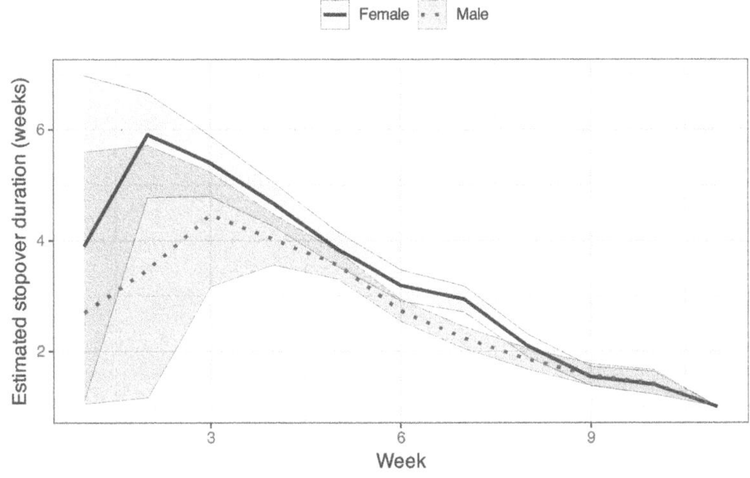

FIGURE 8.3: Stopover duration by sex and arrival week. Means and 95% credible intervals (shaded areas) are provided.

Conclusion

To wrap up this book, let me share a few pieces of advice. This is not rocket science, just some lessons I've picked up along the way from building hidden Markov models and analyzing data with Bayesian statistics.

- Make your ecological question explicit. First things first. Make sure you've spent some to time to make your ecological question explicit. This step will help you to stay on course, and make the right choices. For example, it's fine to use subsets of your data to address different questions.

- Now in terms of modeling. Think of observations and states first. Don't jump on your keyboard right away. Spend some time thinking about your model with pen and paper. In particular make sure you have the observations and the states of your HMM. Then write down the observation and transition matrices on paper. Write down the transition matrix first. You may act as if you had no imperfect detection. This is really what you're after, the ecological process (survival, dispersal, etc.). Then proceed with the observation matrix.

- When it comes to model fitting with NIMBLE, start simple, with all parameters constant for example. Make sure convergence is reached. Then add complexity one step at a time. Time effect for example, or random effects, or uncertainty in the assignment of states.

- Consider doing simulations to better understand your model. When it comes to model building, consider simulating data to better understand your model. You will always learn something on your model by seeing it an engine to generate data, instead of estimating its parameters. The nice thing with NIMBLE is that you can use your model to simulate data.

- Another advice, quite general in programming, is to not try to optimize your code or to try to make it elegant right away. Make your model work first, then think of optimization.

I hope to have convinced you that hidden Markov models combined with the Bayesian framework are very flexible to analyze capture-recapture data. With your data, you may ask a myriad of questions. The limit is your imagination (and CPU time).

To stimulate your imagination, I have assembled a searchable list at https://oliviergimenez.github.io/curated-list-of-HMM-CR-papers/ with HMM analyses of capture-recapture data for you to get inspiration. This list is not exhaustive, please get in touch with me if you'd like to add a reference.

J. Albert and J. Hu. *Probability and Bayesian Modeling*. Chapman and Hall/CRC, 1st edition, 2019.

N. A. Arnason. Parameter estimates from mark-recapture experiments on two populations subject to migration and death. *Researches on Population Ecology*, 13(2):97–113, 1972.

N. A. Arnason. The estimation of population size, migration rates and survival in a stratified population. *Researches on Population Ecology*, 15 (2):1–8, 1973.

P. Besbeas and B.J.T. Morgan. Exact inference for integrated population modelling. *Biometrics*, 75(2):475–484, 2019.

C. Brownie, J. E. Hines, J. D. Nichols, K. H. Pollock, and J. B. Hestbeck. Capture-recapture studies for multiple strata including non-Markovian transitions. *Biometrics*, 49(4):1173–1187, 1993.

S. T. Buckland. A Conversation with Richard M. Cormack. *Statistical Science*, 31(2):142–150, 2016.

R. Choquet, A. Viallefont, L. Rouan, K. Gaanoun, and J.-M. Gaillard. A semi-Markov model to assess reliably survival patterns from birth to death in free-ranging populations. *Methods in Ecology and Evolution*, 2(4):383–389, 2011.

D. Cole. *Parameter Redundancy and Identifiability*. Chapman and Hall/CRC, 2020.

P. B. Conn, D. S. Johnson, P. J. Williams, S. R. Melin, and M. B. Hooten. A guide to Bayesian model checking for ecologists. *Ecological Monographs*, 88(4):526–542, 2018.

P.B. Conn and E.G. Cooch. Multistate capture–recapture analysis under imperfect state observation: An application to disease models. *Journal of Applied Ecology*, 46(2):486–492, 2009.

E.G. Cooch and G. White. *Program MARK: a gentle introduction*. 13th edition, 2017. URL http://www.phidot.org/software/mark/docs/book/.

E.G. Cooch, P.B. Conn, S.P. Ellner, A.P. Dobson, and K.H. Pollock. Disease dynamics in wild populations: modeling and estimation: a review. *Journal of Ornithology*, 152(2):485–509, 2012.

L.M.M Cook, L. P. Brower, and H. J. Croze. The accuracy of a population estimation from multiple recapture data. *Journal of Animal Ecology*, 36(1):57–60, 1967.

P. Couet, F. Gally, C. Canonne, and A. Besnard. Joint estimation of survival and breeding probability in female dolphins and calves with uncertainty in state assignment. *Ecology and Evolution*, 9(23):13043–13055, 2019.

S. Cubaynes, R. Pradel, R. Choquet, C. Duchamp, J.-M. Gaillard, J.-D. Lebreton, E. Marboutin, C. Miquel, A.-M. Reboulet, C. Poillot, P. Taberlet, and O. Gimenez. Importance of accounting for detection heterogeneity when estimating abundance: the case of French wolves. *Conservation Biology*, 24(2):621–626, 2010.

S. Cubaynes, C. Lavergne, E. Marboutin, and O. Gimenez. Assessing individual heterogeneity using model selection criteria: how many mixture components in capture–recapture models? *Methods in Ecology and Evolution*, 3(3):564–573, 2012.

P. de Valpine, D. Turek, C. J. Paciorek, C. Anderson-Bergman, D. T. Lang, and R. Bodik. Programming with models: writing statistical algorithms for general model structures with NIMBLE. *Journal of Computational and Graphical Statistics*, 26(2):403–413, 2017.

J. A. Dupuis. Bayesian estimation of movement and survival probabilities from capture-recapture data. *Biometrika*, 82(4):761–772, 1995.

C. R. Faustino, C. S. Jennelle, V. Connolly, A. K. Davis, E. C. Swarthout, A. A. Dhondt, and E. G. Cooch. infection dynamics in a house finch population: seasonal variation in survival, encounter, and transmission rate. *Journal of Animal Ecology*, 73(4):651–669, 2004.

X. Fernández-i Marín. ggmcmc: Analysis of mcmc samples and Bayesian inference. *Journal of Statistical Software*, 70(9):1–20, 2016.

M. Frederiksen, J.-D. Lebreton, R. Pradel, R. Choquet, and O. Gimenez. Identifying links between vital rates and environment: A toolbox for the applied ecologist. *Journal of Applied Ecology*, 51(1):71–81, 2014.

J. Gabry and T. Mahr. bayesplot: Plotting for Bayesian models, 2022. URL https://mc-stan.org/bayesplot/. R package version 1.10.0.

A. Gelman and J. Hill. *Data Analysis Using Regression and Multilevel/Hierarchical Models*. Cambridge University Press, 2006.

A. Gelman, J. Hwang, and A. Vehtari. Understanding predictive information criteria for Bayesian models. *Statistics and Computing*, 24(6): 997–1016, 2014.

A. Gelman, A. Vehtari, D. Simpson, C. C. Margossian, B. Carpenter, Y. Yao, L. Kennedy, J. Gabry, P.-C. Bürkner, and M. Modrák. Bayesian workflow, 2020.

M. Genovart and R. Pradel. Transience effect in capture-recapture studies: The importance of its biological meaning. *Plos One*, 14(9): e0222241, 2019.

M. Genovart, R. Pradel, and D. Oro. Exploiting uncertain ecological fieldwork data with multi-event capture–recapture modelling: an example with bird sex assignment. *Journal of Animal Ecology*, 81(5): 970–977, 2012.

V. Gervasi, L. Boitani, D. Paetkau, M. Posillico, E. Randi, and P. Ciucci. Estimating survival in the Apennine brown bear accounting for uncertainty in age classification. *Population Ecology*, 59(2):119–130, 2017.

O. Gimenez, V. Rossi, R. Choquet, C. Dehais, B. Doris, H. Varella, J.-P. Vila, and R. Pradel. State-space modelling of data on marked individuals. *Ecological Modelling*, 206(3):431–438, 2007.

O. Gimenez, S. J. Bonner, R. King, R.A. Parker, S.P. Brooks, L.E. Jamieson, V. Grosbois, B.J.T. Morgan, and L. Thomas. WinBUGS for population ecologists: Bayesian modeling using Markov chain Monte Carlo methods. In D. L. Thomson, E. G. Cooch, and M. J. Conroy, editors, *Modeling Demographic Processes in Marked Populations*, pages 883–915. Springer, 2009a.

O. Gimenez, B. J. T. Morgan, and S. P. Brooks. Weak identifiability in models for mark-recapture-recovery data. In D. L. Thomson, E. G. Cooch, and M. J. Conroy, editors, *Modeling Demographic Processes in Marked Populations*, pages 8–48. Springer, 2009b.

O. Gimenez, J.-D. Lebreton, J.-M. Gaillard, R. Choquet, and R. Pradel. Estimating demographic parameters using hidden process dynamic models. *Theoretical Population Biology*, 82(4):307–316, 2012.

O. Gimenez, E. Cam, and J.-M. Gaillard. Individual heterogeneity and capture–recapture models: what, why and how? *Oikos*, 127(5): 664–686, 2018a.

O. Gimenez, J.-D., Lebreton, R., Choquet, R. Pradel, and R2ucare: An R package to perform goodness-of-fit tests for capture–recapture models. *Methods in Ecology and Evolution*, 9(7):1749–1754, 2018b.

B.R. Goldstein, D. Turek, L. Ponisio, and P. de Valpine. nimbleEcology: Distributions for ecological models in 'nimble', 2021. URL https: //CRAN.R-project.org/package=nimbleEcology. R package version 0.4.1.

T. A. Gowan, M. D. Tringali, J. A. Hostetler, J. Martin, L. I. Ward-Geiger, and J. M. Johnson. A hidden Markov model for estimating age–specific survival when age and size are uncertain. *Ecology*, 102(8): e03426, 2021.

V. Grosbois, O. Gimenez, J. M. Gaillard, R. Pradel, C. Barbraud, J. Clobert, A. P. Møller, and H. Weimerskirch. Assessing the impact of climate variation on survival in vertebrate populations. *Biological Reviews*, 83(3):357–399, 2008.

S. Guérin, D. Picard, R. Choquet, and A. Besnard. Advances in methods for estimating stopover duration for migratory species using capture-recapture data. *Ecological Applications*, 27(5):1594–1604, 2017.

J. B. Hestbeck, J. D. Nichols, and R. A. Malecki. Estimates of movement and site fidelity using mark-resight data of wintering Canada geese. *Ecology*, 72(2):523–533, 1991.

D. Jurafsky and J. H. Martin. Appendix on Hidden Markov models. In D. Jurafsky and J. H. Martin, editors, *Speech and Language Processing*. Prentice Hall, 3rd edition, 2023. URL https://web.stanford.edu/~jur afsky/slp3/.

S. Kanyamibwa, A. Schierer, R. Pradel, and J.-D. Lebreton. Changes in adult survival rates in a western European population of the White Stork *ciconia ciconia*. *Ibis*, 132(1):27–35, 1990.

M. Kéry and M. Schaub. *Bayesian Population Analysis Using WinBUGS: a Hierarchical Perspective*. Academic Press, 2011.

R. King, B. J. T. Morgan, O. Gimenez, and S. P. Brooks. *Bayesian Analysis for Population Ecology*. Chapman and Hall/CRC, 2009.

D. N. Koons, M. Gamelon, J.-M. Gaillard, L. M. Aubry, R. F. Rockwell, F. Klein, R. Choquet, and O. Gimenez. Methods for studying cause-specific senescence in the wild. *Methods in Ecology and Evolution*, 5(9): 924–933, 2014.

M. Kéry and K. Kellner. *Applied Statistical Modelling for Ecologists: A Practical Guide to Bayesian and Likelihood Inference Using R, JAGS, NIMBLE, Stan and TMB*. Elsevier, Academic Press, 2024.

J.-D. Lebreton, K. P. Burnham, J. Clobert, and D. R. Anderson. Modeling survival and testing biological hypotheses using marked animals: A unified approach with case studies. *Ecological Monographs*, 62(1): 67–118, 1992.

J.-D. Lebreton, J. D. Nichols, R. J. Barker, R. Pradel, and J. A. Spendelow. Modeling individual animal histories with multistate capture–recapture models. *Advances in Ecological Research*, 41(3):87–173, 2009.

W. A. Link and M. J. Eaton. On thinning of chains in mcmc. *Methods in Ecology and Evolution*, 3(1):112–115, 2012.

L. Marescot, S. Benhaiem, O. Gimenez, H. Hofer, J.-D. Lebreton, X. A. Olarte-Castillo, S. Kramer-Schadt, and M. L. East. Social status mediates the fitness costs of infection with canine distemper virus in Serengeti spotted hyenas. *Functional Ecology*, 32(5):1237–1250, 2018.

M. A. McCarthy. *Bayesian Methods for Ecology*. Cambridge University Press, 2007.

M. A. McCarthy and P. Masters. Profiting from prior information in Bayesian analyses of ecological data. *Journal of Applied Ecology*, 42(6): 1012–1019, 2005.

B. T. McClintock, R. Langrock, O. Gimenez, E. Cam, D. L. Borchers, R. Glennie, and T. A. Patterson. Uncovering ecological state dynamics with hidden Markov models. *Ecology Letters*, 23(12):1878–1903, 2020.

R. McElreath. *Statistical Rethinking: A Bayesian Course with Examples in R and Stan*. Chapman and Hall/CRC, 2nd edition, 2020.

S. B. McGrayne. *The Theory That Would Not Die: How Bayes' Rule Cracked the Enigma Code, Hunted Down Russian Submarines, and Emerged Triumphant from Two Centuries of Controversy*. Yale University Press, 2011.

J. D. Nichols, J. E. Hines, K. H. Pollock, R. L. Hinz, and W. A. Link. Estimating breeding proportions and testing hypotheses about costs of reproduction with capture-recapture data. *Ecology*, 75(7):2052–2065, 1994.

J.D. Nichols, J.R. Sauer, K. H. Pollock, and J. B. Hestbeck. Estimating transition probabilities for stage-based population projection matrices using capture–recapture data. *Ecology*, 73(1):306–312, 1992.

V. Ollivier, R. Choquet, A. Gamble, M. Bastien, B. Combes, E. Gilot-Fromont, M. Pellerin, J.-M. Gaillard, J.-F. Lemaître, H. Verheyden, and T. Boulinier. Temporal dynamics of antibody level against lyme disease bacteria in roe deer: Tale of a sentinel? *Ecology and Evolution*, 13(8):e10414, 2023.

S. Paganin and P. de Valpine. Computational methods for fast Bayesian model assessment via calibrated posterior p-values. *Journal of Computational and Graphical Statistics*, 34(2):462–473, 2025.

L. C. Ponisio, P. de Valpine, N. Michaud, and D. Turek. One size does not fit all: Customizing mcmc methods for hierarchical models using nimble. *Ecology and Evolution*, 10(5):2385–2416, 2020.

R. Pradel. Multievent: An Extension of Multistate Capture–Recapture Models to Uncertain States. *Biometrics*, 61(2):442–447, 2005.

R. Pradel. The stakes of capture–recapture models with state uncertainty. In D. L. Thomson, E. G. Cooch, and M. J. Conroy, editors, *Modeling Demographic Processes in Marked Populations*, pages 781–795. Springer, 2009.

R. Pradel and A. Sanz-Aguilar. Modeling trap-awareness and related phenomena in capture-recapture studies. *Plos One*, 7(3):e32666, 2012.

R. Pradel, C.M.A. Wintrebert, and O. Gimenez. A proposal for a goodness–of–fit test to the Arnason–Schwarz multisite capture–recapture model. *Biometrics*, 59(1):43–53, 2003.

R. Pradel, , O. Gimenez, and J.-D. Lebreton. Principles and interest of gof tests for multistate capture–recapture models. *Animal Biodiversity and Conservation*, 28.2:189–204, 2005.

R. Pradel, L. Maurin-Bernier, O. Gimenez, M. Genovart, R. Choquet, and D. Oro. Estimation of sex-specific survival with uncertainty in sex assessment. *Canadian Journal of Statistics*, 36(1):29–42, 2008.

L.R. Rabiner. A tutorial on hidden Markov models and selected applications in speech recognition. *Proceedings of the IEEE*, 77(2):257–286, 1989.

C.P. Robert and G. Casella. *Monte Carlo Statistical Methods*. Springer, 2nd edition, 2004.

C.P. Robert and G. Casella. *Introducing Monte Carlo Methods with R.* Springer, 2010.

L. Rouan, R. Choquet, and R. Pradel. A general framework for modeling memory in capture—recapture data. *Journal of Agricultural, Biological, and Environmental Statistics*, 14(3):338–355, 2009.

J. A. Royle. Modeling individual effects in the Cormack–Jolly–Seber model: A state–space formulation. *Biometrics*, 64(2):364–370, 2008.

J.A. Royle, R. B. Chandler, R. Sollmann, and B. Gardner. *Spatial Capture-Recapture*. Academic Press, 2013.

C. S. Rushing. An ecologist's introduction to continuous-time multi-state models for capture–recapture data. *Journal of Animal Ecology*, 92(4):936–944, 2023.

S. Santoro, I. Pacios, S. Moreno, A. Bertó-Moran, and C. Rouco. Multi-event capture–recapture modeling of host–pathogen dynamics among European rabbit populations exposed to myxoma and Rabbit Hemorrhagic Disease Viruses: Common and heterogeneous patterns. *Veterinary Research*, 45(1):39, 2014.

C. J. Schwarz, J. F. Schweigert, and A. N. Arnason. Estimating migration rates using tag-recovery data. *Biometrics*, 49:177–193, 1993.

R. P. Scofield, D. J. Fletcher, and C. J. R. Robertson. Titi (sooty shearwaters) on Whero island: Analysis of historic data using modern techniques. *Journal of Agricultural, Biological, and Environmental Statistics*, 6(2):268–280, 2001.

D. Turek. basicmcmcplots: Trace plots, density plots and chain comparisons for mcmc samples, 2022. URL https://mc-stan.org/bayesplot/. R package version 0.2.7.

D. Turek, P. de Valpine, and C. J. Paciorek. Efficient Markov chain Monte Carlo sampling for hierarchical hidden Markov models. *Environmental and Ecological Statistics*, 23(4):549–564, 2016.

D. Turek, C. Wehrhahn, and O. Gimenez. Bayesian non-parametric detection heterogeneity in ecological models. *Environmental and Ecological Statistics*, 28(2):355–381, 2021.

A. Vehtari, A. Gelman, D. Simpson, B. Carpenter, and P.-C. Bürkner. Rank-normalization, folding, and localization: An improved $\hat{R}$ for assessing convergence of MCMC (with discussion). *Bayesian Analysis*, 16(2):667–718, 2021.

B. K. Williams, J. D. Nichols, and M. J. Conroy. *Analysis and Management of Animal Populations*. Academic Press, 2002.

C. Youngflesh. MCMCvis: Tools to visualize, manipulate, and summarize mcmc output. *Journal of Open Source Software*, 3(24):640, 2018.

W. Zucchini, I.L. MacDonald, and R. Langrock. *Hidden Markov models for time series: An introduction using R*. Chapman and Hall/CRC, 2nd edition, 2016.

Index

Note: Locators in *italics* represent figures and **bold** indicate tables in the text.

A

Acceptance probability, 17, 20

ACF plots, *see* Autocorrelation function plots

Additive effect, 173

Age- and cause-specific mortality

model and NIMBLE implementation, 295–299

motivation, 294–295

results and interpretation, 299, *299*

Age-specific breeding probabilities

model and NIMBLE implementation, 300–307

motivation, 299–300

results and interpretation, 307, **308**

Age uncertainty

model and NIMBLE implementation, 255–259

motivation, 254–255

results and interpretation, 259–261, **260**

AIC, *see* Akaike Information Criterion

Akaike Information Criterion (AIC), 148, 149

Alive–dead models, 133

Bayesian motivation

moment matching, 189–190

prior elicitation, 188–189

capture–recapture data, 134–136, *135*, *136*

Cormack–Jolly–Seber (CJS) model, 133–134

derivatives, 142–148

fitting (NIMBLE), 136–142

covariates, 154

individual covariates, 168–172

multiple covariates, 173–178

random effects, 179–182

temporal covariates, 155–168

time-varying individual covariates, 182–187

goodness of fit, 150

classical tests, 151–154

design considerations, 154

posterior predictive checks, 150

Watanabe Information Criterion (WAIC), model comparison with, 148–150
Apparent survival, 153, 154
Arnason, Neil, 193
Arnason-Schwarz (AS) model, 193; *see also* Sites and states models
  fitting
    Geese data, 196–197
    goodness of fit, 207–208
    NIMBLE implementation, 198–207
    for states, 220
    theory, 193–196
AS model, *see* Arnason-Schwarz model
Autocorrelation, 26, 28, 29
Autocorrelation function (ACF) plots, 28
Average likelihood, 6, 9

B
Bayes, Thomas, 3
Bayesian approach, 5–7, 8, 149, 229
  with MCMC methods, 96
Bayesian formulation, 84–86
Bayesian inference, 15, 246; *see also* Prior distribution
  incorporating prior information
    moment matching, 189–190
    prior elicitation, 188–189
Bayesian statistics, 3
  assessing convergence, 23
    burn-in period, 23–26
    chain length, 26–29
    convergence issues, 29–30
  Markov chain Monte Carlo (MCMC), 12, *12*
    Markov chains, 14–15
    Metropolis algorithm, 15–23
    Monte Carlo integration, 13–14
  posterior approximation via numerical integration, 7–12, *10*
Bayes' theorem, 3–5, 6, 7, 8, 9, 11, 12, 16
Bernoulli distribution, 84
Beta distribution, 10–11, *10*, 13, 86, 190, 209
BGR statistic, *see* Brooks-Gelman-Rubin statistic
Binomial distribution, 7, 10, 51, 52, 75, 78
Binomial likelihood, 8, 10, 34, 78
Brooks-Gelman-Rubin (BGR) statistic, 25
Brute-force approach, 102–104, 106
BUGS language, 33
BUGS pieces of software, 149
`buildMCMC()`, 54, 58, 76
Burn-in period, 23–26, 38, 180

C
`calculate()`, 67
Candidate value, 16–17, 18, 20, 26, 62, 64

Capture–recapture data, 91–93,
114, 123, 133, 134–136, *135*,
136, 150, 151, 245, 294, 316
Categorical distribution, 84, 85,
87, 100
Centered parameterization, 181
Chain length, 26–29
`character()`, 48
CJS model, *see*
Cormack–Jolly–Seber
model
`coda::effectiveSize()` function,
29
Compilation, in NIMBLE, 67–68
`compileNimble()`, 54, 56, 57, 58, 76
Complete likelihood, 95–96, 102,
110, 118, 127, 128
Conditional independence, 105
Conditional probabilities, 3, 81,
82, 83, 84, 95, 105
`configureMCMC()`, 54, 58, 61, 67, 76
`configureRJ()` function, 248, 249
Convergence, assessing, 23, 43
burn-in period, 23–26
chain length, 26–29
convergence issues, 29–30
Cormack, Richard, 134
Cormack–Jolly–Seber (CJS)
model, 133–134, 247
derivatives, 142–148
fitting (NIMBLE), 136–142
Covariates, 154, 245
age, uncertainty in
model and NIMBLE
implementation, 255–259
motivation, 254–255

results and interpretation,
259–261, **260**
individual covariates, 168
continuous, 169–172
discrete, 168–169
multiple covariates, 173–178
random effects, 179–182
reversible jump MCMC
model and NIMBLE
implementation, 247–250
motivation, 245–247
results and interpretation,
250–254
sex, uncertainty in
model and NIMBLE
implementation, 261–265
motivation, 261
results and interpretation,
265–267, **266**
temporal covariates
continuous, 164–168
discrete, 155–164
time-varying individual
covariates, 182–187
Custom implementation, 107–111;
see also NIMBLE

D
DAG, *see* Directed acyclic graph
`dbern()`, 88
`dcat()`, 87, 88
`dCJSxx()`, 146
Declarative language for building
models, 35
Decoding after marginalization,
118

average first, compute after, 126, *127*, *128*
compute first, average after, 124–125
implementation, 121–124
theory, 118–120, *121*
Detection probability, 11, 93, 134, 141, 148, 153, 212, 232, 247, 258, 260, 266, 269, 283
Deviance Information Criterion (DIC), 148–149
dHMM() function, 203, 275
dHMMhomemade function, 114
dHMMo function, 111, 112, 118
DIC, *see* Deviance Information Criterion
Directed acyclic graph (DAG), 36
Dirichlet prior, 209–214, 209, 216, 275
Discrete random variable, 5
Dupuis, Jérôme, 226

E
equals(), 256

F
Folk theorem of statistical computing, 30
forecast::ggAcf() function, 28
Forward algorithm, 104–107, 118, 120

G
Gamelon, Marlène, 295
Geese data, 196–197, 203, 208, 273
Gelman, Andrew, 30
Genovart, Meritxell, 262

Gervasi, Vincenzo, 255
getViterbi() function, 123, 125
Goodness of fit of CJS model, 150, 207–208
classical tests, 151–154
design considerations, 154
posterior predictive checks, 150

H
Hamiltonian Monte Carlo, 33
Hidden Markov models (HMMs), 78, 91, 94–95, 194, 195, 200, 261, 269, 275, 280, 295, 300, 309, 315; *see also* Markov models
Bayesian formulation, 84–86
capture-recapture data, 91–93
decoding after marginalization, 118
average first, compute after, 126, *127*, *128*
compute first, average after, 124–125
implementation, 121–124
theory, 118–120, *121*
likelihood, 95–96
longitudinal data, 78–79
marginalization, 102
brute-force approach, 102–104
forward algorithm, 104–107
NIMBLE implementation, 107–113
model fitting with NIMBLE, 96–101

NIMBLE implementation,
86–91
observation matrix, 93–94
pooled encounter histories,
113–118
HMMs, *see* Hidden Markov
models

I

Indexing, in NIMBLE, 67
Individual covariates, 168
continuous, 169–172
discrete, 168–169
Individual heterogeneity, 179, 225,
268
model and NIMBLE
implementation,
280–292, **285**
motivation, 280
results and interpretation,
292, **292**
Initialization, incomplete and
incorrect (NIMBLE),
73–75
`integer()`, 48
`integrate()`, 9, 11
Inverse-logit function, 158, 163,
253
Inverse probability, 4

J

JAGS, 33, 45, 67, 71
Jolly, George, 134

L

Laplace, Pierre-Simon, 3
Laplace approximation, 33

Lebreton, Jean-Dominique, 247
Life history traits, quantifying,
294
age- and cause-specific
mortality
model and NIMBLE
implementation, 295–299
motivation, 294–295
results and interpretation,
299, *299*
age-specific breeding
model and NIMBLE
implementation,
300–307
motivation, 299–300
results and interpretation,
307, **308**
stopover duration
model and NIMBLE
implementation, 309–312
motivation, 309
results and interpretation,
312–313, *313*
Likelihood, 6, 81–83, 95–96, 98,
114, 137, 200, 236; *see also*
Marginalization
average, 9
binomial, 8, 10, 34, 78
complete, 95–96, 102, 110, 118,
127, 128
marginal, 96, 102, 104, 106,
112, 113, 118, 146
maximum likelihood
estimation (MLE), 5, 6, 8,
226, 227, 228
pooled, 118

Local minima, issue of, 226–230, *228*
`logical()`, 48
Logit function, 158
Log-likelihood function, 15, 64
Longitudinal data, 78–79
  Markov model for, 79
    assumptions, 80
    example, 84
    initial states, 81
    likelihood, 81–83
    transition matrix, 80–81

M
Marginal distribution, 12, 209
Marginalization, 102; *see also*
    Likelihood
  brute-force approach,
    102–104
  decoding after, 118
    average first, compute after,
    126, *127*, *128*
    compute first, average after,
    124–125
    implementation, 121–124
    theory, 118–120, *121*
  forward algorithm, 104–107
  NIMBLE implementation
    custom implementation,
    107–111
    `nimbleEcology` package,
    111–113
Marginal likelihood, 96, 102, 104,
    106, 112, 113, 118, 146
Marginal probability, 4
Markov chain Monte Carlo
    (MCMC), 12, *12*

Metropolis algorithm, 15–23
Monte Carlo integration,
    13–14
NIMBLE, MCMC samplers in
    default samplers, 61–62
    updating MCMC chains,
    68–69
    user-defined samplers,
    62–66
Markov chains, 14–15, 19, 23, 25,
    26, 30, 85, 95
Markovian property, 82, 94, 207
Markov models; *see also* Hidden
    Markov models (HMMs)
  for longitudinal data, 79
    assumptions, 80
    example, 84
    initial states, 81
    likelihood, 81–83
    transition matrix, 80–81
  memory effects in
    model and NIMBLE
      implementation, 273–278
    motivation, 272–273
    results and interpretation,
      278–280
Markov process, 80, 81, 91, 286
Marzolin, Gilbert, 134, 136
Maximum likelihood estimation
    (MLE), 5, 6, 8, 226, 227,
    228
MCMC, *see* Markov chain Monte
    Carlo
`MCMCplot()` function, 42
`MCMCsummary()` function, 41
`MCMCtrace()` function, 42
Memory effects in Markov models

model and NIMBLE
implementation, 273–278
motivation, 272–273
results and interpretation,
278–280, **279**
Memory model, 273
Metropolis algorithm, 15–23, 64
Metropolis-Hastings algorithm,
16
Mixing, 27, 28–29, 30, 110, 141, 181
MLE, *see* Maximum likelihood
estimation
`model$calculate()`, 56
Model lack of fit, 268
individual heterogeneity,
accommodating
model and NIMBLE
implementation,
280–292, **285**
motivation, 280
results and interpretation,
292–293, **292**
memory effects in Markov
models
model and NIMBLE
implementation, 273–278
motivation, 272–273
results and interpretation,
278–280, **279**
trap-dependence, accounting
for
model and NIMBLE
implementation, 269–272
motivation, 268–269
results and interpretation,
272
Moment matching, 189–190

Monte Carlo Expectation
Maximization, 33
Monte Carlo integration, 13–14
Multinomial logit, 209, 214–217
Multiple covariates, 173–178
`myfunction()`, 49
`my_metropolis()`, 64, 65

N
NIMBLE, 32–33, *32*, 84
faster compilation, 67–68
getting started, 34–47
implementation, 86–91,
198–207, 220–225,
255–259, 261–265,
280–292
custom implementation,
107–111
`nimbleEcology` package,
111–113
incomplete and incorrect
initialization, 73–75
indexing, 67
internal workflow, 54–61
logo of, *33*
MCMC chains, updating,
68–69
MCMC samplers in
default samplers, 61–62
user-defined samplers,
62–66
model fitting with, 96–101
parallelization, 71–73
precision vs. standard
deviation, 67
programming in, 47

calling R/C++ functions,
49–51
functions, 47–49
user-defined distributions,
51–53
reproducibility, 69–70
vectorization, 75–76
`nimbleCode()` function, 34
`nimbleEcology` package, 111–113
`nimbleExternalCall()`, 50
`nimbleMCMC()`, 38–39, 45, 54, 59, 61,
69, 70, 73, 76, 149
`nimbleModel()`, 54, 67, 73, 76
`nimbleRcall()`, 50
Non-centered parameterization,
181
Non-homogeneous geometric
process, 312
Numerical integration, posterior
approximation via, 7–12,
10

O
Observation matrix, 93–94, 111,
195, 200, 220, 222, 232,
235, 237, 240, 258, 259,
263, 264, 265, 271, 273,
275, 282, 298, 305–306,
311, 315
OpenBUGS, 33
`optim()` function, 226

P
Parallelization, 71–73
`parLapply()`, 72
Pooled encounter histories,
113–118

pooled likelihood, 118
Posterior approximation via
numerical integration,
7–12, *10*
Posterior density function, 15–16
Posterior distribution, 6–7, 10, 11,
12, 13, 15, 32, 37, 38, 39, 41,
42, 44, 45, 59, 124, 149, 167,
171, 175, 229, 230, 312
Pradel, Roger, 262
Precision vs. standard deviation,
67
Predictive accuracy, 148, 150
Price, Richard, 3
Prior distribution, 6, 10, 15–16,
189, 214; *see also* Bayesian
inference
Probability distribution, 5
Proposal distribution, 16, 17, 26,
28, 29, 64, 65, 248, 249

R
`R1ucare` package, 151
Random effects, 179–182
Random variable, 5, 8, 13, 35, 85,
212
`rcat()` function, 84
Reproducibility, MCMC, 69–70
Reversible jump MCMC
(RJMCMC), 245, 246–247
model and NIMBLE
implementation, 247–250
motivation, 245–247
results and interpretation,
250–254
`rHMMhomemade` function, 115
Richdale, Lance, 217

RJMCMC, *see* Reversible jump
    MCMC
Rtools, 34
runMCMC(), 54, 58, 59, 69, 76

S
sample() function, 85
samplesSummary(), 59
Sanz-Aguilar, Ana, 269
Schwarz, Carl, 193
Scofield, Richard, 219
Seber, George, 134
Second-order Markov process,
    207, 273
Sex uncertainty
    model and NIMBLE
        implementation, 261–265
    motivation, 261
    results and interpretation,
        265–267, **266**
Sites and states models, 193; *see
    also* Arnason-Schwarz
    (AS) model
    local minima, issue of,
        226–230, *228*
    multiple sites, 208
        Dirichlet prior, 209–214,
            209, *213*
        multinomial logit, 214–217
    uncertainty, 230
        breeding states, 230–237
        disease states, 237–241, *238*
States as sites, 217
    Arnason-Schwarz (AS) model
        for states, 220
    NIMBLE implementation,
        220–225

titis data, 217–219, *218*
stopCluster(), 72
Stopover duration
    model and NIMBLE
        implementation, 309–312
    motivation, 309
    results and interpretation,
        312–313, *313*
survival$calculate(), 55

T
Temporal covariates
    continuous, 164–168
    discrete, 155–164
Time-effect parameterization,
    309
Time-varying individual
    covariates, 182–187
Titis data, 217–219, *218*, 231, 242
Trace plot, 19, 21, 22, 23, 24, 25, 26,
    28, 230
Transition matrix, 14–15, 80–81,
    85, 86, 111, 112, 145, 156,
    195, 199, 211, 215, 220, 221,
    232, 234, 239, 240, 256,
    262, 270, 273, 274,
    281–282, 286, 296, 301,
    303, 310
Trap-dependence, 151, 153–154,
    207
    model and NIMBLE
        implementation, 269–272
    motivation, 268–269
    results and interpretation,
        272
Turek, Daniel, 286

U

Uncertainty, 230
    in age
        model and NIMBLE
           implementation, 255–259
        motivation, 254–255
        results and interpretation,
           259–261, **260**
    breeding states, 230–237
    disease states, 237–241, *238*
    in sex
        model and NIMBLE
           implementation, 261–265
        motivation, 261
        results and interpretation,
           265–267, **266**

V

Vague priors, 8
Vectorization, NIMBLE, 75–76
`viterbi()` function, 123

Viterbi algorithm, 118, 119, 120,
    121, *121*, 123, 124–125, 128;
    see also Decoding after
    marginalization

W

WAIC, *see* Watanabe Information
    Criterion
Watanabe Information Criterion
    (WAIC), 148, 224
    model comparison with,
        148–150
WBWA tests, *see* "Where before
    where after" tests
"Where before where after"
    (WBWA) tests, 207–208
Widely Applicable Information
    Criterion, 148
WinBUGS, 33, 45
`workflow()` function, 72

X

Xcode, 34

For Product Safety Concerns and Information please contact our EU
representative GPSR@taylorandfrancis.com
Taylor & Francis Verlag GmbH, Kaufingerstraße 24, 80331 München, Germany